Evaluating Sustainable Development and Corporate Social Responsibility Projects

Tony Kealy

Evaluating Sustainable Development and Corporate Social Responsibility Projects

An Ethnographic Approach

 Springer

Tony Kealy 🄳
School of Electrical and Electronic Engineering
Technological University Dublin, City Campus
Dublin, Ireland

ISBN 978-3-030-38675-7   ISBN 978-3-030-38673-3   (eBook)
https://doi.org/10.1007/978-3-030-38673-3

This Springer imprint is published by the registered company Springer Nature Switzerland AG
The registered company address is: Gewerbestrasse 11, 6330 Cham, Switzerland

Dedicated to my wife Olivia, my son Luke, my daughter Emily, and in memory of our son Tom

Contents

Abbreviations

AC	Alternating Current
ARR	Accounting Rate of Return
BCG	Boston Consultancy Group
BES	Business Expansion Scheme
BMS	Building Management System
BS	British Standard
CC	Corporate Citizenship
CCGT	Combined-Cycle-Gas-Turbine
CEO	Chief Executive Officer
CER	Commission for Energy Regulation
CF	Capacity Factor
CFO	Chief Financial Officer
CO_2	Carbon Dioxide
COV	Coefficient Of Variation
CR	Corporate Responsibility
CSD/T	Catholic Social Doctrine/Teaching
CSO	Corporate Social Opportunity
CSR	Corporate Social Responsibility
CT	Current Transformer
DC	Direct Current
DCENR	Department of Communications, Energy, and Natural Resources
DCF	Discounted Cash Flow
DFIG	Double-Fed-Induction-Generator
DIT	Dublin Institute of Technology
DJEI	Department of Jobs, Enterprise and Innovation
DM	Digital Meter
DSM	Demand Side Management
DSO	Distribution System Operator
EGIP	Embedded Generation Interface Protection
EIIS	Employment and Investment Incentive Scheme

EMC	Electro Magnetic Compatibility
EPA	Environmental Protection Agency
EPSSU	Energy Policy Statistical Support Unit
ESB	Electricity Supply Board
EU	European Union
FERC	Future Earnings Response Coefficient
FIT	Feed-In-Tariff
GAA	Gaelic Athletic Association
GHG	Green-House-Gas
GRI	Global Reporting Initiative
GRLI	Global Responsible Leadership Initiative
ICTU	Irish Congress of Trade Unions
IDA	Industrial Development Authority
IEA	International Energy Agency
IFA	Irish Farmers Association
IGBT	Insulated Gate Bipolar Transistor
ILO	International Labour Organisation
IRR	Internal Rate of Return
IS	Irish Standard
ISO	Irish Standards Organisation
KPI	Key Performance Index
kVA	kilo-Volt-Amps
kW	kilo-Watt
kWh	kilo-Watt-hour
LCA	Life Cycle Assessment
M. Phil.	Master of Philosophy
MC	Measure of Commitment failure
MIC	Maximum Import Capacity
MNC	Multi-National Corporation
MNE	Multi-National Enterprise
MVA	Mega Volt Amp
MW	Mega Watt
NDP	National Development Plan
NGO	Non-Governmental-Organisation
NPV	Net Present Value
OECD	Organisation for Economic Cooperation and Development
PC	Personal Computer
PFC	Power Factor Correction
PG Dip	Post-Graduate Diploma
PLC	Public Limited Company
PP	Payback Period
PR	Public Relations
PSO	Public Service Obligation
PV	Photo-Voltaic
RDM	Rotating-Disc-Meter

RPM	Rotations Per Minute
RSPM	Relative Sustainable Performance Measure
SBD	Sustainable Business Development
SD	Sustainable Development
SDG	Sustainable Development Goals
SEAI	Sustainable Energy Authority of Ireland
SEM	Single Electricity Market
SEMO	Single Electricity Market Operator
SME	Small-Medium-Enterprise
SMP	System Marginal Price
SP	Simple Payback
SPSS	Statistical Package for Social Science
SSM	Supply Side Management
SWA	Steel-Wire-Armour
TBL	Triple Bottom Line
TMT	Top Management Team
UNGC	United Nations Global Compact

List of Figures

List of Tables

Chapter 1
Introduction

Abstract This discusses the background to the sustainable development philosophy with a specific focus on the role of wind turbines in the efforts to decarbonise the economy. There is a global drive towards a 'green economy'. Traditional electricity generation has focused on the burning of fossil fuels to power the majority of the generating plant. Wind turbine technology is expected to reduce our dependence on fossil-fuel-driven electrical generators. This expected fossil-fuel reduction lessens the amount of harmful Green-House Gases, mainly CO_2, emitted into the atmosphere. One of the weaknesses in the sustainable development sphere is identified, namely, the methods by which the effectiveness of wind turbines is measured. This study aims to develop an evaluation framework to address this weakness. A step-by-step guide as to how the framework is developed is described in this chapter.

1.1 Background to This Research

1.1.1 Sustainable Business Development

Business activity and the economic strength of a nation are inextricably linked (OECD Economic Surveys 2018). A potential for a business to be in a position to generate economic activity, provide income for a workforce, and support a community presents many challenges, as well as many opportunities (Buhmann et al. 2019; Greenwood and Freeman 2018). With these corporate challenges come significant responsibilities. It is generally accepted that the primary duties of any business are to treat all its employees with dignity and respect, to provide fair wages, and carry out operations efficiently and effectively without degrading natural or environmental resources while at the same time generating profit (Piscicelli et al. 2018). Demands for accountability and transparency in the public domain are steadily increasing. Enhanced communications and global connectivity have put the spotlight on Sustainable Business Development (SBD) and its interrelated term Corporate Social Responsibility (CSR). CSR is seen as a vehicle for achieving sustainable development (Weber et al. 2014), and the two terms are used interchangeably in this book.

© Springer Nature Switzerland AG 2020

T. Kealy, *Evaluating Sustainable Development and Corporate Social Responsibility Projects*, https://doi.org/10.1007/978-3-030-38673-3_1

The global connectivity means that irresponsible or unethical business practices can now be communicated to a vast audience at the touch of a button. One such example is the Volkswagen (VW) car manufacturers' inaccurate CO_2 emission claims which cost the company on reputation and financial levels (Hotten 2015). SBD and CSR are not new; the first primary academic concept of CSR was presented by Bowen in the 1950s (Bowen 1953). The CSR theme later evolved into the business ethics domain in the 1980s (Carroll 1999). Despite the CSR concept being around for several decades, there appears to be a lack of literature purporting a cohesive SBD/CSR approach (Pour et al. 2014). Elkington (1997) did propose a Triple-Bottom-Line (TBL) model which encourages organisations to consider, not just their well-established financial bottom line, but also their environmental and social bottom lines. Some researchers (Tate and Bals 2018) claim that while the financial and environmental dimensions have been considered comprehensively by management theory and practice, the social dimension remains fundamentally under-represented. SBD (and its shortened version Sustainable Development, SD), CSR, and TBL are used interchangeably in the literature (Silberhorn and Warren 2007). The TBL concept has evolved from its initial idea as a framework that allows for the reporting on the financial, environmental, and social dimensions in business to a synonym for sustainability (Isil and Hernke 2017). The SD/CSR/TBL area covers a multitude of interwoven business issues and therefore it is desirable for individuals who are active in the area to possess a multidisciplinary perspective (Isil and Hernke 2017). This author has qualifications and experience in both technical engineering and business management fields and seeks to address some of the issues raised by Isil and Hernke (2017). An ethnographic emphasis is adopted when carrying out on-site case studies, on-site action research, semi-structured interviews and an online survey into sustainable development. The outputs of these different methodologies/methods allow for the development of a new, unique and original evaluation framework. The novel framework can be used to evaluate and enhance sustainable development strategic business decisions concerning business investing in wind turbine initiatives in the context of environmental protection (to mitigate climate change and global warming). Climate policy decision-making is challenging because it involves the assessment of climate data that covers a multitude of scientific fields and disciplines but also must support the collection of knowledge from many stakeholders referred to as group decision-making (Nikas et al. 2018). This current study embraces stakeholder theory (Miles 2017; Carroll 1999) which considers a broad range of people or groups that affect, or are affected by, the activities and operations of a business organisation.

The book is written with no prior hidden agendas and has very much an independent focus. Given its current topical focus, the researcher was cognizant of the need to ensure independence and impartiality throughout the study process with no vested interest represented. The closed-loop framework and the independent, robust wind turbine data presented in this book offer an opportunity to enhance Elkington's (1997) TBL model and assist decision-making in the area of wind turbine investments. Such valuable data is essential to accurately critique a perceived social consensus, that all wind turbine projects contribute to the reduction in CO_2 emissions and also helps to make significant savings for the investors. This consensus of perceived benefits of wind turbine technology is tested in this study. Objective measured data is used in the decision evaluation process. Corbett et al. (2018) argue that it is difficult to determine the environmental impact of sustainability projects a priori, i.e. based on theoretical deduction rather than empirical observation. The closed-loop systematic framework and associated empirical wind turbine data presented in this book assist decision-makers in making more informed evidence-based decisions.

1.1.2 Drive Towards a 'Green Economy' and CO_2 Emission Reduction

An International Energy Agency (IEA) report published in October 2019 claimed that global offshore wind capacity is set to increase 15-fold over the next two decades, turning it into a \$1 trillion business (IEA 2019, p. 3). While the case studies presented in this book are onshore wind turbines, it is clear that global energy agencies predict that wind can play a major part in the battle to reduce CO_2 emissions. One of the primary applications for wind turbines is in the electricity generation sector and is the focus of the turbine applications considered in this book. More work needs to be done in this area as, despite impressive gains for renewables, fossil fuels still account for nearly two-thirds of electricity generation globally. Global energy-related CO_2 emissions reached a historic high in 2018, driven by an increase in coal use in the power sector (IEA 2019, p. 11). One of the main output variables being analysed in this current book is CO_2 emissions. CO_2 emission (reduction) is one of the main variables by which the strategic decision-making regarding wind turbine investments are evaluated. Wind turbines electrical generators are embedded in parallel with fossil-fuel electrical generators and can be used locally (on-site) or fed directly into the national electricity grid. It appears reasonable to expect to see a reduction in on-site CO_2 emissions when a business utilises renewable energy sources such as wind turbines to provide non-fossil-fuel-generated electrical power to their premises via the national electricity grid (Al-Masri et al. 2018). Renewable energy sources are mainly classed as carbon-neutral sources of energy (Gils and Simon 2017), i.e. they release zero carbon per kWh of energy produced. The electrical energy produced by such renewable sources 'offsets' a similar electrical energy unit that would have been produced by conventional (fossil-fuel) electrical generators (Cullen 2013). The closed-loop control theory adopted in this research was initially used in the electrical and mechanical domains (Sagawa et al. 2017), but the principles can also be applied to the business domain. A closed-loop feedback system is one where deviations between what the output variables are meant to be doing and what they are actually doing are diagnosed (Sagawa et al. 2017), and if there is an error between the two (desired and measured performance) then a correction step is usually applied to mitigate the error. The opposite of closed-loop control is open-loop control which does not use feedback, the actions are taken, and it is hoped and expected that the output variable is doing what it is intended to do. A 'Life Cycle Assessment' (Yang et al. 2018) of wind turbines is not included in this research study. The initial manufacturing stage (raw materials used) and end-of-life stages (disposal/recycling) are not considered; instead, the operational stage in the production of kWh energy units by the wind turbines is solely used to evaluate the wind turbine investment decisions. The carbon intensity of electricity benchmark value is expected to decrease as more renewable energy sources (such as wind) become operational on the national electricity grid. The carbon intensity of a given fuel is the amount of carbon dioxide that is released by burning that fuel to produce one-unit (kWh) of energy and is measured in (k)g CO_2/kWh. For example, the carbon intensity benchmark for electricity generation in 2016 was 482.8 g CO_2/kWh (Sustainable Energy Authority Ireland 2017; Table A4-1). The 2016 value increased from a carbon intensity value of 467.5 g CO_2/kWh in 2015. The increase was due to a 23% increase in the gas generation to compensate for lower wind (less windy than 2015) and hydro generation (lower rainfall) and a switch from net electricity imports to electricity exports to the UK. The 2015 value (467.5 g CO_2/kWh) had also increased by 2.5% on the previous year (456.6 g CO_2/kWh in 2014) mainly as a result of the 19.6% increase in coal used for electricity generation (Sustainable Energy Authority Ireland 2017). For most fossil fuels the carbon intensity value is almost constant, but in the case of electricity generation it depends on the fuel mix used to generate the electricity and on the efficiency of the technology employed. The carbon intensity benchmark (g CO_2/kWh) for electricity generation in each year as reported by the Sustainable Energy Authority of Ireland is an average energy mix value and is used in the renewable electricity generation 'offset' carbon emission metrics. As the yearly average energy mix values vary from year to year, there may be different carbon intensity benchmark values for electricity generation used throughout this book depending on the year being assessed (Table A4-1). Multiplying the electricity production offset by the annual average emissions rate (carbon intensity) value indicates the number of emissions offset by wind power (Cullen 2013).

It is important to note that, despite the focus in this book on wind turbine applications being utilised as renewable electrical generators embedded with fossil-fuel electrical generators, there are other potential uses for wind turbine generators. Wind turbine generators could be used to produce low-carbon hydrogen. Hydrogen is the most common fuel used in fuel cells. One of the methods used to produce hydrogen is by the electrolysis technique. The electrolysis technique can make use of wind turbines as its energy source. Hydrogen can be stored in many ways and is therefore ideally suited to the intermittent and fluctuating phenomenon associated with wind turbine power outputs. The storability also enhances decentralised electricity generation (off-grid generation), where the embedding process is deemed to be problematic (Ayodele and Munda 2019). The coupled operation of wind turbine and hydrogen electrolyser can provide a zero-carbon fuel that could be used to provide power system flexibility (Sarrias-Mena et al. 2015). It can also be used in sectors that are hard to decarbonise, including industry, refining, and transport (IEA 2019, p. 45). Stand-alone wind turbines can be specifically designed to produce local, renewable-based hydrogen that can be used in the iron, steel and chemical industries. Also, the output from the dedicated turbines could be used in transport where fuel cell electric vehicles powered by hydrogen would be well suited to heavy trucks, aviation, and shipping (IEA 2019, p. 55). The IEA (2019) report also claims that a 1-gigawatt wind project could produce enough low-carbon hydrogen gas to heat approximately 250,000 homes (IEA 2019, p. 14).

This research is carried out in line with Irish and international efforts to confront stated objectives and principles. In an Irish context, the national strategy for public investment in research is aimed at improving the economic and societal well-being of the country, and the health and well-being of its citizens (Forfás 2012). This Forfás (2012) report recommends 14 priority research areas which should receive most of the public research funding. An action plan was developed for each of the 14 research areas which ranged from Future Networks & Communications (priority area A) to Innovation in Services and Business Processes (priority area N). Other research areas include medical devices and manufacturing competitiveness. This research study is aligned to priority area K which concentrates on Smart Grids & Smart Cities—using technologies and design solutions to more effectively and efficiently manage complex infrastructure systems, enable greater resource efficiency and help move to a low-carbon society. Another important stakeholder in the environmental area in Ireland is the Environmental Protection Agency (EPA). The role of the EPA is to ensure that Ireland's environment is protected by monitoring changes and detecting early warning signs of neglect or deterioration. The EPA is an independent public body, and they work with many different organisations which have specific environmental functions and responsibilities for issues such as air quality, energy, environmental education, green schools, and waste disposal among others.

1.1.2.1 Irelands National Plan on CSR

In April 2014, the Irish Minister for Jobs, Enterprise and Innovation Mr Richard Bruton T. D. launched 'Good for Business, Good for the Community', Ireland's (first) National Plan on Corporate Social Responsibility 2014–2016. The National Plan on CSR pronounces a vision for Ireland to be recognised as a centre of excellence for responsible and sustainable business practices. The measures contained in the plan are

- Establish a Stakeholder Forum on CSR to support the development of CSR in Ireland,
- Establish a baseline of CSR activity in Ireland, through the National Standards Association of Ireland,
- Work with stakeholders to raise awareness of CSR and support best practice CSR,
- Explore how the Industrial Development Authority (IDA) and Enterprise Ireland can promote CSR with their client companies, and
- Support programmes to develop CSR in the Small–Medium Enterprises (SME) sector.

In the follow-up CSR plan entitled 'Towards Responsible Business' (2017) the Irish government claimed that the main objective of the first plan was to raise awareness of the benefits of CSR to businesses and all stakeholders in society. The second national CSR (2017) plan aims to build on the progress and focus on several specific areas for action. The Irish national policies are aligned with international CSR frameworks and the development of the procedures are cognisant of the principles outlined by a number of organisations among which are the European Union (EU) Commission, the United Nations Global Compact (July 2000), the Organisation for Economic Co-operation and Development Guidelines for Multinational Enterprises (OECD 1976), the International Labour Organisation Tripartite Declaration of Principles concerning Multinational Enterprises and Social Policy (ILO 2000), and ISO 26000 guidance standard on social responsibility by the Department of Jobs, Enterprise and Innovation (DJEI 2014).

The environmental pillar of the DJEI (2014) national plan states that key drivers in the 'green economy' include emission reduction targets, renewable energy targets (for electricity, transport, and heating), energy efficiency targets, increasing fossil-fuel prices, environmental legislation, and consumer preferences. This study seeks to determine if indeed Ireland has become a centre of excellence for CSR practices as proposed in the Irish national plan and if the contribution of wind turbines as renewable electrical energy sources makes a positive contribution to the CO_2 emission reduction targets.

1.1.2.2 Horizon 2020

From a European research perspective, Horizon 2020 reflects the policy priorities of the Europe 2020 strategy and addresses significant challenges faced by citizens in Europe and elsewhere. Funding focuses on the following societal challenges:

- Health, demographic change and well-being,
- Food security, sustainable agriculture and forestry, marine and maritime and inland water research, and the bioeconomy,
- Secure, clean and efficient energy,
- Smart, green and integrated transport,
- Climate action, environment, resource efficiency and raw materials,
- Europe in a changing world—inclusive, innovative and reflective societies, and
- Secure societies—protecting freedom and security of Europe and its citizens.

The *secure, clean and efficient energy* challenge is concerned with reducing energy consumption and carbon emissions by, among other things, employing alternative energy sources. The Horizon 2020 framework states that this challenge is underpinned by 'robust decision-making and public engagement'. The *climate action, environment, resource efficiency, and raw materials* challenge is focused on activities that help to keep the average global warming below 2 °C. This current research study focuses on progress made in two specific challenges, namely *secure, clean and efficient energy,* and *climate action, environment, resource efficiency and raw materials*, with a particular focus upon the Irish context. The Irish government collaborates with Horizon 2020, and in doing so aims to secure €1.25bn research and innovation funding from the European framework. The Irish societal challenges are addressed in the Innovation 2020, Ireland's Strategy for Research and Development, Science and Technology with a vision that Ireland can become a Global Innovation Leader driving a strong, sustainable economy and a better society.

1.1.3 Irish Energy Benchmarks

Progress towards Ireland becoming a leader in sustainability and a better society with particular emphasis on all aspects of energy is overseen by the Sustainable Energy Authority Ireland (SEAI) with assistance from its specialist statistics team the Energy Policy Statistical Support Unit (EPSSU). Each year the SEAI produces a report on all aspects of energy in Ireland and the report is used in meeting its international reporting obligations, as well as advising policymakers and informing investment decisions (SEAI 2016, 2017). The SEAI 'Energy in Ireland' report states, among other things, Ireland's energy import dependency percentage benchmarks. Some historical energy import dependency benchmark values are 85% for 2012, 90% for 2013, 85% for 2014, 88% for 2015 and 69% in 2016. The much-improved energy import dependency value in 2016 came about because of the Corrib gas field coming on stream. Energy-related CO_2 emissions increased by 3.6% in 2016. One of the striking values in the SEAI (2017, p. 20) report is that 52% of the 4812 ktoe energy inputs are lost in both 'Own Use/Transmission Loss' (254 ktoe) and 'Electricity Transformation Loss' (2242 ktoe). It was anticipated that the increasing use of Combined Cycle Gas Turbine (CCGT) plants such as Tynagh (384-MW) in 2006 and Huntstown 2 in 2007 (401-MW) would have significantly helped to improve the electricity supply efficiency. However, the year/percentage loss energy benchmarks, as shown in the first and last columns in Table 1.2 do not validate this expectation and surely warrant more intense scrutiny (Kealy 2019). The efficiency is defined as the final consumption of electricity divided by the fuel inputs required to generate this electricity and expressed as a percentage. As more renewable energy sources, particularly wind, are added to the national electricity grid, the efficiency of the system should increase. Wind is termed a direct electricity input and does not have the transformation losses associated with fossil fuels and combustible renewables (SEAI 2017, p. 21). By the end of 2016, the installed capacity of wind generation reached 2827-MW (SEAI 2017, p. 34). There are 1119-MW of wind generation contracted for the connection before the end of 2017 and a further 1458-MW by the end of 2018. These values suggest that wind is a significant contributor to the electricity generation mix. As an example of the overall instantaneous magnitude of the system demand in Ireland, the actual system demand for 11 January 2017 (highest value for that day) was 4558-MW at 17.45 (Eirgrid 2017). Another reason why it is surprising that the energy losses are so high is that ESB Networks have spent €5.695 billion capital expenditure on its infrastructure between 2006 and 2016 (ESB Networks 2017) as shown in Table 1.1.

Table 1.1 Capital
expenditure of ESB Networks
on infrastructure

Year	Capital expenditure
2006	€620,000,000
2007	€600,000,000
2008	€630,000,000
2009	€600,000,000
2010	€604,000,000
2011	€510,000,000
2012	€395,000,000
2013	€421,000,000
2014	€448,000,000
2015	€494,000,000
2016	€373,000,000
Total	€5,695,000,000

Taking these three issues, i.e. the addition to the electricity generation mix of a significant number of wind turbines, the increased use of CCGT technology and the electricity grid upgrade it would appear reasonable to expect significantly lower percentage loss benchmark values than that shown in the final column in Table 1.2.

This book reviews the data provided by such government publications (SEAI 2016, 2017) and the aim is to develop a closed-loop framework in order to provide for a robust evaluation of the decisions to invest in wind turbines, and the possible

Table 1.2 Year/percentage loss energy benchmarks for electricity generation

Year	Primary energy input (ktoe)	Own use/transmission loss (ktoe)	Electricity transformation loss (ktoe)	Percentage loss (%)
2016	4812	254	2242	52
2015	4499	245	2046	51
2014	4365	262	1960	51
2013	4382	262	2004	52
2012	4622	270	2244	54
2011	4506	272	2097	53
2010	4925	282	2445	55
2009	4840	281	2402	55
2008	5102	303	2518	55
2007	5043	229	2550	55
2006	5116	303	2690	59
2005	5100	280	2726	59

contribution of wind energy in the battle against climate change and global warming. The proposed novel framework should be of interest to many of the stakeholders involved in wind turbine initiatives, namely shareholders, employees, customer suppliers, communities, governmental bodies and political groups.

1.2 Aim and Objectives of This Study

This book aims to develop a novel closed-loop framework that can be used by businesses to evaluate decisions to invest in wind turbine initiatives as part of their sustainable development/CSR efforts. The framework in this study concentrates on the environmental component of the three generally accepted components of Sustainable Development, namely *human, financial, and environmental* components (*people, profit, and planet*). Within the broad environmental pillar of sustainable development, this research study concentrates specifically on the climate change aspect of the environmental element and investigates the contribution that an alternative energy source such as wind turbines may or may not make in the ongoing battle against climate change and global warming.

Therefore, the key research aim of this study is

- To develop an analytical closed-loop framework that can be used by businesses to critically evaluate their decision to invest in SD/CSR initiatives specifically concerning wind turbine projects undertaken by the business (**Aim 1**).

The objectives of the research study are to

- Review and critique relevant published literature to identify gaps (weaknesses) in the SD/CSR/Wind Turbine space (**Obj. 1**),
- Evaluate technical aspects of wind turbine installations (**Obj. 2**),
- Carry out economic assessments on wind turbine investments (**Obj. 3**),
- Identify key enablers and inhibitors that have the potential to influence the integration of SD/CSR and strategy within an organisation (**Obj. 4**).

1.3 Methodology

1.3.1 Philosophical Underpinning of the Research

Crotty (1998) stated that the philosophical beliefs and assumptions of the researcher shape the understanding of a research aim, the methods used in a study, and how findings are interpreted. 'Research paradigm', a term coined by Kuhn (1962), denotes a particular world view of the researcher that is influenced by his/her values, beliefs, and methodological assumptions (Locke and Golden-Biddle 2002). O'Neil and Koekemoer (2016), Gaus (2017) advise researchers to state their philosophical assumptions.

Two fundamental philosophical paradigms are positivist and interpretive approaches. A positivist management research approach is based on empirical social science methods with an emphasis on validity, reliability, and generalisations (McInnes et al. 2017) while the interpretive management research philosophy is characterised by a belief in a socially constructed, subjectively-based reality, one that is influenced by culture and values (Packard 2017). The interpretive paradigm allows the researcher to uncover, interpret and understand meanings in a phenomenon within a group or individual context (Matabooe et al. 2016). A subset of the interpretive paradigm utilised in this research is the 'phenomenological' approach which seeks to understand the world through direct experience of a phenomenon. The research philosophy adopted for this research study belongs primarily to the interpretive paradigm.

The book is primarily an ethnographic study with data acquired using surveys, interviews, questionnaires, and case studies. The author examined the phenomenon and interacted with the participants in their real-life, natural environment. This perspective included both the social interactions, behaviours, and perceptions of human participants and (renewable energy) processes operating within the context of their normal working operations.

1.3.2   Methodologies Used in This Research Study

Several different research methodologies were used in this study, namely the ethnographic, survey, action research, and case study methodologies. A range of methodologies was needed as the development of different stages of the framework called for different methodologies to be applied at each specific stage. The ethnographic qualitative research methodology was one of the primary methodologies utilised in the writing of this book. The ethnographic approach provided rich insights into the participant's views, opinions and actions about renewable energy as part of the discussion on sustainable development projects. The survey methodology (study the sampling of individual units from a target population, Callegaro et al. 2015) utilised Thematic analysis as the method of analysis. Thematic analysis is a method for identifying, analysing, and reporting themes within the data (Braun and Clarke 2006). Thematic analysis is a means of analysing qualitative data in a rigorous and methodical manner (Nowell et al. 2017). Thematic analysis was undertaken by developing codes in the initial analysis of interview data presented in Chap. 3 and the survey data presented in Chap. 7. Themes derived from the codes were developed based on the rich data obtained as a result of the wide-ranging organisations which took part in the project. Themes were interpreted to theorise the significance of the patterns and their broader meanings and implications (Braun and Clarke 2006). The action research (Zhang et al. 2015; Dick et al. 2015) methodology was used in Chaps. 4, 5, and 6. The first publication in the area of action research was by Lewin (1946) entitled 'Action Research and Minority Problems'. The action research methodology was a good fit for the researcher in this study as it offered the opportunity to bridge the divide between research and practice (Zhang et al. 2015). Action research requires the researcher to engage with the personnel in the participant business where the research is taking place and subsequently satisfy both their research interests *and* the companies' need for a wind turbine project evaluation. Chaps. 4, 5, and 6 are also examples of the case study methodology. A case study is an empirical enquiry that investigates a phenomenon within its real-life context and where multiple sources of evidence are used (Yin 2008) such as the electric utility bills, annual production output, and sustainability reports as evidence utilised in Chaps. 4, 5, and 6. Yin (2008) also offers a robust guide to address four tests for case study research design, namely construct validity, internal validity, external validity, and reliability.

1.3.3 Research Design

The function of a good research design is to (i) detail the actions to be undertaken to complete the study and (ii) ensure that the validity and objectivity of the study are not compromised (Totawar and Prasad 2016). The real-world flexible research design for this study has its foundation in the critique and review of the literature published in the CSR area. This focus provides the basis from which all the subsequent research decisions are made. A review of the literature identified a lack of a coherent framework (Xu et al. 2016) that could be used by stakeholders in this vital area of business activity. Much of the published research appeared to be rhetorical rather than robust empirical studies. The action research (Zhang et al. 2015; Dick et al. 2015) design used in part of this study (the objective measurement stage) was cyclical and designed around an initial preliminary, pilot exploratory study conducted in the area of CSR/Environment/Wind Turbine. The first publication in the field of action research was by Lewin (1946) entitled 'Action Research and Minority Problems'. Lewin was a German psychologist who focused on ways of participating in social issues to address conflicts, crises, and change and contended that complex real events could not be investigated under laboratory conditions (De Villiers et al. 2007). Part of the action research methodology implementation for this study included the researcher completing the SOLAS 'Safe Pass' training programme which allowed access to the various test sites. The survey methodology utilised in Chap. 7 followed the rigorous qualitative research procedures to ensure the validity and reliability of the study. These five procedures are as follows: collecting and organising the data—coding the data—identifying themes—interpreting the data—presenting the results. The mixed-method approach incorporates several research tools employed to triangulate and corroborate findings enabling new perspectives on a topic to be developed (Brown et al. 2017). This evaluation research study was carried out over many years and is referred to as a longitudinal study. The broad range of experiences/qualifications of this researcher helps to dovetail the academic and work environments.

1.4   Step-by-Step Development of the Framework

The book is structured around three critical parameters as the framework is developed, namely (i) reviewing wind turbine literature through a sustainable development lens, (ii) an empirical technical and economic evaluation of operational wind turbine projects, and (iii) identifying the key enabling and inhibiting factors in managing the sustainable development process. The initial review of the three interconnected constituents that are generally considered to contribute to sustainable development decision-making within a business was identified, namely People, Profit, and Planet (Senechal 2017). This is the basis for the TBL model proposed by Elkington (1997). The new framework is developed iteratively but is based on an initial request for an evaluation of a decision to invest in a 10-kW three-phase wind turbine initiative by a private investor. During the evaluation process for the 10-kW wind turbine project, it was identified that despite much literature purporting to contribute to evaluate sustainability initiatives using the TBL framework, the empirical measurement and reporting dimension of wind turbine installations had not been developed entirely using the TBL framework. This is even though the Environment (Planet) dimension of sustainability to which wind turbines belong (i) appears to be gaining media and political attention (Connolly 2017) and (ii) there is much literature in the area, but much of the writing appears to be rhetorical. The climate change aspect of the environmental component of sustainable development is very topical at present being the main focus of many commentators (Ambrose 2017) and the private investor of the 10-kW wind turbine (presented in Chap. 4) made the decision to invest in such a project to contribute in the fight against climate change. Much literature claims that energy-related CO_2 emissions are the primary source of Green-House Gases that are causing climate change and global warming (Wang et al. 2018). Within this climate change realm, it should be noted that there are a number of strategies that could be employed in order to mitigate the human effect on climate change, namely better control on pollutants (and a slowdown in the phenomenon of the throw-away culture) and also a more efficient method of usage of electrical power (Demand-Side Management, Aghajani et al. 2017). However, the framework for this study is developed using objective measured data from wind turbine projects (Chaps. 4, 5, and 6) as an alternative energy system (Supply-Side Management, Karunanithi et al. 2017) with the aim of reducing CO_2 emissions generally associated with fossil-fuel-driven electrical generators, an area of specific interest to this researcher. The alternative energy sources available include Biomass, Solar Photo Voltaic (PV), Hydroelectric, and Wind Turbines. This latter option (Wind Turbines) is the alternative energy system utilised in the development of the decision-making evaluation framework for this book. The experience of the author allows for many data collection and analysis techniques to be employed in the objective 'Data Measurement' stage in the development of the framework (Stage 3 in Fig. 8.3). Rigour is applied to this vital data measurement stage by using a data triangulation method. After this data triangulated measurement stage, an assessment of the issues that appear to enable or inhibit effective sustainability strategy and overall business strategy integration within an organisation were identified. The output of each of these individual stages contributes to the development of a comprehensive closed-loop framework and provides wind energy feedback information into the wind energy literature and the senior decision-making level of the organisation where the wind turbine investment decision is evaluated. The evidenced-based feedback information produced using this new evaluation framework contributes to negate the possible poor senior-level decision-making as a result of uncertain information (Kelman et al. 2017). One of the main components of the new framework and a significant contribution from this study is the introduction of the objective *data measurement* stage. This development stage may help to shift the impetus from sustainable development being a subjective topic into the realm of becoming an objective, measurable concept. The new framework is an extension of Elkington's (1997) model with a focus on the empirical measurement and reporting of pertinent wind turbine data.

References

Aghajani, G. R., Shayanfar, H. A., & Shayeghi, H. (2017). Demand side management in a smart micro-grid in the presence of renewable generation and demand response. *Energy, 126,* 622–637. https://doi.org/10.1016/j.energy.2017.03.051.

Al-Masri, H. M. K., AbuElrub, A., & Ehsani, M. (2018). *Optimisation and layout of a wind farm connected to a power distribution system.* In IEEE International Conference on Industrial Technology 2018 (ICIT) (pp. 1049–1054).

Ambrose, J. (2017, June 30). Shell boss first of 'big oil' executives to call for financial transparency on climate change. *The Daily Telegraph,* p. 4. London, Business, Friday, Edition 1, National Edition.

Ayodele, T. R., & Munda, J. L. (2019). Potential and economic viability of green hydrogen production by water electrolysis using wind energy resources in South Africa. *International Journal of Hydrogen Energy, 44*(33), 17669–17687. https://doi.org/10.1016/j.ijhydene.2019.05.077.

Bowen, H. R. (1953). *Social responsibilities of the businessman.* New York, NY: Harper & Row.

Braun, V., & Clarke, V. (2006). Using thematic analysis in psychology. *Qualitative Research in Psychology, 3*(2), 77–101.

Brown, G., Strickland-Munro, J., Kobryn, H., & Moore, S. A. (2017). Mixed methods participatory GIS: An evaluation of the validity of qualitative and quantitative mapping methods. *Applied Geography, 79,* 153–166. https://doi.org/10.1016/j.apgeog.2016.12.015.

Buhmann, K., Jonsson, J., & Fisker, M. (2019). Do no harm and do more good too: Connecting the SDGs with business and human rights and political CSR theory. *Corporate Governance, 19*(3), 389–403. https://doi.org/10.1108/CG-01-2018-0030.

Callegaro, M., Manfreda, K. L., & Vehovar, V. (2015) *Web survey methodology* (p. 318). London: Sage. ISBN 9780857028617.

Carroll, A. B. (1999). Corporate social responsibility: Evolution of a definitional construct. *Business and Society, 38*(3), 268–295.

Connolly, P. (2017, June 25). Winds of change; as demand for a slice of the renewable energy boom soars, developers face turbulent times. *The Sunday Times,* p. 5. London, Business & Money, Business, Edition 1, Ireland.

Corbett, J., Webster, J., & Jenkin, T. A. (2018). Unmasking corporate sustainability at a project level: Exploring the influence of institutional logics and individual agency. *Journal of Business Ethics, 147*(2), 261–286. https://doi.org/10.1007/s10551-015-2945-1.

Crotty, M. (1998). *The foundations of social research: Meaning and perspective in the research process*. London: Sage. ISBN: 9780761961062.

Cullen, J. (2013). Measuring the environmental benefits of wind-generated electricity. *American Economic Journal: Economic Policy, 5*(4), 107–133. https://doi.org/10.1257/pol.5.4.107.

De Villiers, M. R., Lubbe, S., & Klopper, R. (2007). Action research: The participative researcher or experiential approach. *Alternation, 14*(1), 218–242.

Department of Jobs, Enterprise, and Innovation (DJEI). (2014). *"Good for business, good for the community", Ireland national plan for corporate social responsibility 2014–2016*. https://www.djei.ie/. Accessed on 2 March 2017.

Dick, B., Sankaran, S., Shaw, K., Kelly, J., Soar, J., Davies, A., et al. (2015). Value co-creation with stakeholders using action research as a meta-methodology in a funded research project. *Project Management Journal, 46*(2), 36–46. https://doi.org/10.1002/pmj.2483.

Eirgrid. (2017). Available at http://www.eirgridgroup.com/how-the-grid-works/system-information/. Accessed on 13 June 2017.

Elkington, J. (1997). *Cannibal with forks: The triple bottom line of 21st century business*. Gabriola Island, BC: Capstone.

ESB Networks. (2017). *Memories from ESB archives*. Every Annual Report, 1928–2016. https://esbarchives.ie/2016/02/17/esb-annual-reports/. Accessed on 11 July 2017.

Forfás. (2012). https://www.djei.ie/en/National-Research-Prioritisation-Exercise-First-Progress-Report.pdf. Accessed on 27 February 2017.

Gaus, N. (2017). Selecting research approaches and research designs: A reflective essay. *Qualitative Research Journal, 17*(2), 99–112. https://doi.org/10.1108/QRJ-07-2016-0041.

Gils, H. C., & Simon, S. (2017, February). Carbon neutral archipelago—100% renewable energy supply for the Canary Islands. *Applied Energy, 188*, 342–355.

Greenwood, M., & Freeman, R. E. (2018). Deepening ethical analysis in business ethics. *Journal of Business Ethics, 147*(1), 1–4. https://doi.org/10.1007/s10551-017-3766-1.

Hotten, R. (2015, December 10). *Volkswagen: The scandal explained*. BBC News. Available at http://www.bbc.com/news/business-34324772. Accessed on 3 June 2017.

IEA. (2019). *Offshore wind outlook 2019*. World Energy Outlook Special Report, International Energy Agency. Available at https://www.iea.org/newsroom/news/2019/october/offshore-wind-to-become-a-1-trillion-industry.html. Accessed on 29 October 2019.

ILO. (2000). *Tripartite declaration of principles concerning multinational enterprises and social policy (MNE Declaration)*—(3rd ed). Available at https://www.ilo.org/empent/Publications/WCMS_101234/lang--en/index.htm. Accessed 4 Jan 2020.

Isil, O., & Hernke, M. T. (2017). The triple-bottom-line: A critical review from a transdisciplinary perspective. *Business Strategy and the Environment, 26*, 1235–1251. https://doi.org/10.1002/bse.1982.

Karunanithi, K., Saravanan, S., Prabakar, B. R., Kannan, S., & Thangaraj, C. (2017). Integration of demand and supply side management strategies in generation expansion planning. *Renewable and Sustainable Energy Reviews, 73*, 966–982. https://doi.org/10.1016/j.rser.2017.01.017.

Kealy, T. (2019). A review of CO_2 emission reductions due to wind turbines using energy benchmarks: A focus on the Irish electrical energy market. *International Journal of Global Warming, 19*(3), 267–292. https://doi.org/10.1504/IJGW.2019.103727.

Kelman, S., Sanders, R., & Pandit, G. (2017). "Tell it like it is": Decision making, groupthink, and decisiveness among US federal subcabinet executives. *Governance: An International Journal of Policy, Administration, and Institutions, 30*(2), 245–261. https://doi.org/10.1111/gove.12200.

Kuhn, T. S. (1962). *The structure of scientific revolutions*. Chicago IL: University of Chicago Press.

Lewin, K. (1946). Action research and minority problems. *Journal of Social Issues, 2*(4), 34–46.

Locke, K., & Golden-Biddle, K. (2002). An introduction to qualitative research: Its potential for industrial and organisational psychology. In S. G. Rogelberg (Ed.), *Handbook of research methods in industrial and organisational psychology* (pp. 99–118). Oxford, UK: Blackwell.

Matabooe, M. J., Venter, E., & Rootman, C. (2016). Understanding relational conditions necessary for effective mentoring of black-owned small businesses: A South African perspective. *Acta Commercii, 16*(1), 1–11. https://doi.org/10.4102/ac.v16i1.327.

McInnes, S., Peters, K., Bonney, A., & Halcombe, E. (2017). An exemplar of naturalistic inquiry in general practice research. *Nurse Researcher, 24*(3), 36–41. https://doi.org/10.7748/nr.2017. e1509.

Miles, S. (2017). Stakeholder theory classification: A theoretical and empirical evaluation of definitions. *Journal of Business Ethics, 142*(3), 71–98. https://doi.org/10.1007/s10551-015-2741-y.

Nikas, A., Doukas, H., & Martinez Lopez, L. (2018, March). A group decision making tool for assessing climate policy risks against multiple criteria. *Heliyon, 4*(3). https://doi.org/10.1016/j. heliyon.2018.e00588.

Nowell, L. S., Norris, J. M., White, D. E., & Moules, N. J. (2017). Thematic analysis: Striving to meet the trustworthiness criteria. *International Journal of Qualitative Methods, 16*, 1–13. https:// doi.org/10.1177/1609406917733847.

OECD. (1976). OECD Guidelines for Multinational Enterprises. Available at http://www.oecd.org/ corporate/mne/1922428.pdf. Accessed on 2 March 2017.

OECD Economic Surveys. (2018, January 1). Maintaining a successful business sector in a changing world, *2018*(1), 57–98, ISSN: 0376-6438.

O'Neil, S., & Koekemoer, E. (2016). Two decades of qualitative research in psychology, industrial and organisational psychology and human resource management within South Africa: A critical review. *SA Journal of Industrial Psychology, 42*(1). https://doi.org/10.4102/sajip.v42i1.1350.

Packard, M. D. (2017). Where did interpretivism go in the theory of entrepreneurship? *Journal of Business Venturing, 32*(5), 536–549. https://doi.org/10.1016/j.busvent.2017.05.004.

Piscicelli, L., Ludden, G. D. S., & Cooper, T. (2018). What makes a sustainable business model successful? An empirical comparison of two peer-to-peer goods-sharing platforms. *Journal of Cleaner Production, 172*, 4580–4591. https://doi.org/10.1016/j.jclepro.2017.08.170.

Pour, B. Saeidi, Nazari, K., & Emami, M. (2014). Corporate social responsibility: A literature review. *African Journal of Business Management, 8*(7), 228–234. https://doi.org/10.5897/AJBM12.106.

Sagawa, J. K., Nagano, M. S., & Neto, M. S. (2017). A closed-loop model of a multi-station and multi-product manufacturing system using bond graphs and hybrid controllers. *European Journal of Operational Research, 258*(2), 677–691.

Sarrias-Mena, R., Fernandez-Ramirez, L. M., Garcia-Vazquez, C. A., & Jurado, F. (2015). Electrolyser models for hydrogen production from wind energy systems. *International Journal of Hydrogen Energy, 40*(7), 2927–2938. https://doi.org/10.1016/j.ijhydene.2014.12.125.

SEAI. (2016, August). *Renewable electricity in Ireland 2015, 2016 report*. Prepared by the Energy Policy Statistical Support Unit. http://www.seai.ie. Accessed 2 November 2016.

Senechal, O. (2017, January). Research directions for integrating the triple bottom line in maintenance dashboards. *Journal of Cleaner Production, 142*(Part 1), 331–342. https://doi.org/10.1016/ j.jclepro.2016.07.132.

Silberhorn, D., & Warren, R. C. (2007). Defining corporate social responsibility: A view from big companies in Germany and the UK. *European Business Review, 19*(5), 352–372. https://doi.org/ 10.1108/09565340710818950.

Sustainable Energy Authority Ireland. (2017, December). *Energy in Ireland 1990–2016* (2017 Report). Available at https://www.seai.ie/resources/publications/Energy-in-Ireland-1990-2016-Full-report.pdf. Accessed on 12 December 2017.

Tate, W. L., & Bals, L. (2018). Achieving shared triple bottom line (TBL) value creation: Toward a social resource-based view (SRBV). *Journal of Business Ethics, 152*(3), 803–826. https://doi. org/10.1007/s10551-016-3344-y.

Totawar, A., & Prasad, M. (2016, July–September). Research methodology: A step-by-step guide for beginners. *South Asian Journal of Management, 23*(3), 210–213, ISBN 978-81-321-0648-7 (PB) [Book review by Ranjit Kumar].

Wang, H., Ang, B. W., & Zhou, P. (2018). Decomposing aggregate CO2 emission changes with heterogeneity: An extended production-theoretical approach. *Energy Journal, 39*(1), 59–79. https://doi.org/10.5547/01956574.39.1.hwan.

Weber, O., Diaz, M., & Schwegler, R. (2014). Corporate social responsibility of the financial sector—Strengths, weaknesses and the impact on sustainable development. *Sustainable Development, 22*(5), 321–335. https://doi.org/10.1002/sd.1543.

Xu, J., Li, L., & Zheng, B. (2016). Wind energy generation technological paradigm diffusion. *Renewable and Sustainable Energy Reviews, 59,* 436–449.

Yang, J., Chang, Y., Zhang, L., Hao, Y., Yan, Q., & Wang, C. (2018). The life-cycle energy and environmental emissions of a typical offshore wind farm in China. *Journal of Cleaner Production, 180,* 316–324. https://doi.org/10.1016/j.jclepro.2018.01.082.

Yin, R. K. (2008). *Case study research: Design and methods* (4th ed.). London: Sage.

Zhang, W., Levenson, A., & Crossley, C. (2015). Move your research from the ivory tower to the board room: A primer on action research for academics, consultants, and business executives. *Human Resource Management, 54*(1), 151–174.

Chapter 2
Corporate Social Responsibility Through a Wind Turbine Lens—A Literature Review

Abstract Corporate Social Responsibility (CSR) remains an ambiguous concept. Reasons for this ambiguity include (i) it crosses into several academic disciplines, (ii) it is not adequately taught in educational institutions, (iii) globalisation has challenged the roles and responsibilities of employees, corporations, and the state. Despite the ambiguity perception, this extensive literature review found that CSR and its related concept Sustainable Development (SD) is generally considered to be concerned with business decision-making that considers economic, social, and environmental factors. Of these three distinct, yet interconnected, factors this study found that there is a substantial body of research assessing the relationship between CSR and financial (economic) business performance and CSR and human (social) aspects but less so assessing the CSR/environmental connection. Research on this understudied aspect is limited, and the literature significantly appears to indicate a lack of integrity or universality, specifically concerning wind turbine projects. Some studies claim that this may be due to the lack of a systematic analytical framework.

2.1 Structure of This Literature Review

The basis for the selection of studies for this literature review utilises the bibliometric approach (De Rezende et al. 2018). This approach measures the usage of the CSR concept by the number of publications in the area (Danilovic et al. 2015). While this method is easy to implement, there is a risk that some less cited articles may have been omitted, some of which may be capable of valuable contributions in their own right. Section 2.1, this current section, details the contents of each section in the review. Section 2.2 presents an overview of the literature in the general area of CSR. It reviews literature from the five perspectives of business management under which most of the CSR studies are published. This chapter also discusses the ambiguous/subjective nature of CSR/Sustainable Development research. Section 2.3 examines literature purporting to identify the responsibilities of many stakeholders who influence the CSR space. Section 2.4 reviews the literature concerning the reporting of sustainability issues particularly in the Triple-Bottom-Line dimensions that are generally considered in CSR activity, namely, economic, environmental, and social aspects.

© Springer Nature Switzerland AG 2020

T. Kealy, *Evaluating Sustainable Development and Corporate Social Responsibility Projects*, https://doi.org/10.1007/978-3-030-38673-3_2

The environmental reporting subsection (Section 2.4.3) includes literature on climate change and alternative energy systems (wind turbines). Section 2.5 critiques literature concerning the critical stakeholders involved in the wind industry from an Irish and global perspective. A brief overview of the other renewable energy technologies is also discussed in this chapter. Section 2.6 presents the results of the review and identifies gaps in the literature that are addressed in this current study.

2.2 Corporate Social Responsibility

2.2.1 Overview

The raising of awareness that corporations have concerning social responsibility has, in part, been driven by changing customer expectations, more regulation, and stakeholder pressure (Hansen et al. 2014). The broad-ranging stakeholder theory appears to have replaced the narrow-focused classical shareholder theory. Shareholder theory holds that the purpose of a business is to provide a return of investment for shareholders and increase the wealth of investors who risk capital in the business. While there are many definitions (Hack et al. 2014) for CSR, the term can be broadly defined as a moral/ethical decision-making ethos adopted by companies which allow their business to contribute to the welfare of society beyond their self-interests (Saeidi et al. 2014). Holme and Watts (2000) declare '*Corporate Social Responsibility is the continuing commitment by businesses to behave ethically and contribute to economic development while improving the quality of life of the workforce and their families as well as of the local community and society at large*'. Bowens' seminal work on CSR in 1953 entitled 'Social Responsibilities of the Businessman' (Bowen 1953) discussed the relationship between businesses and society (Carroll 1999). It laid the foundation for practitioners and academics to consider CSR becoming an integral part of strategic planning and decision-making by business managers. Davis (1960) declared that businesses could not ignore their social responsibilities; however, Friedman (1970) argued that the only responsibility of business was to increase their profits. Friedman (1970) famously stated that the only social responsibility of business is to '*use its resources and engage in activities designed to increase its profits as long as it stays within the rules of the game*'. There have been other shareholder theory protagonists since then (Coors and Winegarden 2005) who question the role of CSR in business activity, but the majority of businesses appear to have embraced the fact that measuring their success on a single bottom line of profit alone does not sustain the company into the future (MacCormac and Haney 2012). A threefold focus called the 'Triple-Bottom-Line' (TBL) on people, planet, and profit was proposed by Elkington (1997). However, de Vries et al. (2015) recently claimed that companies are superficially terming their activities as green, a process termed 'greenwashing' to appear environmentally friendly, while Glavas and Mish (2015) claimed that businesses that adopt a CSR ethos are more responsive ecologically and socially while prospering economically. It is clear from the literature that it is not detrimental for corporations to focus on making a profit to be sustainable, but it must balance its profit-driven agenda in unison with its social responsibilities (Rivera et al. 2017). The design and implementation of CSR can be based on different ethical foundations and motivations (Schaltegger and Burritt 2018). Schaltegger and Burritt (2018) assess four different ethical management versions of CSR, namely, reactionary, reputational, responsible, and collaborative motivation. The last two versions (responsible and collaborative) are the desired motivations for effective CSR management and corporate sustainability. Corporate ethics may have developed rapidly over the past couple of decades since the 1980s, but questions remain as to the effectiveness of corporate ethics to foster ethical behaviour (Mercier and Deslandes 2017). It may be that as Fremeaux and Michelson (2017) claim businesses whose purpose is not exclusively economic have the potential to contribute much more to the common good and that corporate financial needs should be balanced by employees' human needs (Kinley 2017). The 'common good' is described by Mea and Sims (2019) as the 'people first principle' whereby every human being has transcendent value that resides within his or her essence. The ultimate purpose of a business is to serve human needs (Mea and Sims 2019). The motivation for a business operating in a socially responsible and ethical manner include (a) their moral compass, (b) they are doing so because their competitors are doing it, and (c) they are afraid that unethical behaviour is uncovered. Though motives may be questionable (Kearns 2017), increasingly corporations report not solely on well-established financial performance metrics but also report on human and environmental effects of their business operations. While there is an increase in the number of companies that report on sustainability activities within their corporations some researchers (Higgins et al. 2018) argue that sustainability reporting can occur because managers submit to social pressure that makes reporting 'required', 'expected', or 'normal' rather than using it to lead to meaningful change. This current publication assesses the effectiveness of sustainability reporting because some authors (Talbot and Boiral 2018) question the validity of some of the corporate sustainability reports whereby data is manipulated to improve the company's image. Xu et al. (2016) highlight the lack of a cohesive approach, not just to sustainability reporting, but to the many aspects of CSR and the ambiguity therein.

Renewed interest in CSR and its related concept of sustainable development is possible due to technology advances and instantaneous communication channels. CSR can be viewed as a business approach to sustainable development, whereby businesses voluntarily consider environmental, social, and economic issues in their business strategies (Dahlsrud 2008). Stakeholder theory is developed for a long-term perspective in Sustainable Development (Krisnawati et al. 2014). Indeed, Hansen et al. (2014) found that sustainability management and CSR have also become tightly integrated. Bolis et al. (2014) suggested that even though Sustainable Development and CSR are distinct segments, they are strongly related and can be used interchangeably. A review of CSR literature by Wang (2015) claimed that current understanding of CSR and its related terms such as Corporate Citizenship (CC) and sustainable development had been conceptualised within the discipline of management. Management as a discipline refers to that branch of knowledge associated with the principles and practices of basic administration (Okyhusen et al. 2015) to manage resources efficiently and effectively (Matten and Crane 2005). CSR scholars have investigated the CSR themes from five main business management perspectives, as follows:

- *Marketing* (limiting the scope of social obligations to one stakeholder only, namely, consumers, as businesses seek to build a brand image),
- *Leadership* (the ability of business leaders to build relationships with many stakeholders to advance an adequate shared business vision),
- *Strategy* (businesses align their behaviours with their primary stakeholders, embedding CSR into their corporate strategy),
- *Stakeholder theory* (wide-ranging scope of companies to include all stakeholders in the decision-making process),
- *Social obligation* (businesses encompass all their activities with the objectives and values of our society in mind).

Each of the management perspectives is now individually reviewed.

2.2.1.1   Marketing Perspective

The CSR literature from a marketing perspective specifies the narrow focus of the marketing dimension only, focusing on building brands as opposed to the effect on other stakeholders. Mishra and Modi (2016) analysed the link between marketing capability and CSR shareholder wealth. Their study revealed that CSR efforts such as a business using clean energy (environment aspect) make a positive contribution to the image of the company. A case study on an Australian bank (Devin 2016) revealed that while the bank was correct in claiming to be a leading renewable energy financier, it omitted to state that it was also Australia's leading financier of coal mining seen as a significant contributor to global warming. The term used to describe communication by the bank was half-truths (Devin 2016). This study claimed that while organisations keen to be perceived in good Public Relations (PR) light publish their CSR efforts for all the public to see usually in the form of a positive sustainability report, the omission of crucial information may challenge the authenticity of such claims (Devin 2016). The term 'greenwashing' has been used in CSR literature to describe such dubious claims of some organisations to pose as good corporate citizens (Mahoney et al. 2013).

2.2.1.2   Leadership Perspective

Some authors (Maak et al. 2016) attempt to link CSR with a theoretical leadership approach entitled 'responsible leadership'. Maak et al. (2016) refine the concept of responsible leadership to explain how the CSR engagement of organisations is influenced by responsible leadership styles based on perceived moral obligations towards shareholders or stakeholders. Kurucz et al. (2017) explored the role of leadership in implementing strategic sustainability initiatives within an organisation. This complex role requires leaders to establish social, environmental, and economic principles that, when observed in practice, co-creates meanings in sustainability that integrate various stakeholder values and interests. Such a 'relational' leadership style (Kurucz et al. 2017) considers the companies' impact on nature and society. Mea and Sims (2019) argue that profit, trust, and sustainability are the natural outcomes for business leaders who seek to develop virtuous habits. However, not all leaders in organisations participate or can participate in virtuous leadership behaviour, Schyns and Schilling (2013) note that an increasing number of studies investigate different forms of what is termed 'destructive leadership'. As leaders in their company, they rely on followers following them on the path towards company goals. Not surprisingly, most of the outcomes of destructive leadership are assessed from a followers' point-of-view. Schyns and Schilling (2013) found that followers of destructive leaders are likely to have negative attitudes towards the leader and show resistance towards him/her perhaps resulting in counterproductive work tactics. Destructive leaders find it hard to convince followers to 'follow' and are likely to have a negative attitude to the organisation.

2.2.1.3 Strategy Perspective

Battaglia et al. (2016) carried out a longitudinal 8-year study of an Italian food co-operative to analyse how CSR could be integrated with the organisational strategy. The three main instruments used to promote CSR integration were the sustainability report, annual sustainability plan, and participatory social plan. The authors found that there were substantial cognitive barriers that gradually stifled the cognitive enablers to the integration process and suggested that the problems may have arisen as a result of the negative economic performance of the co-operative during the study period. Research by Maas et al. (2016) found that although the integration process between business strategy and sustainability management has been sporadically discussed and conceptualised, there is limited evidence that corporations have successfully implemented the integration.

2.2.1.4 Stakeholder Theory Perspective

The CSR/stakeholder theory link has received much research focus (Ramlugun and Raboute 2015; Demir et al. 2016; Agudo-Valiente et al. 2015). While stakeholder theory is widely accepted the term 'stakeholder' is a contested concept (Miles 2017). The question of 'What is a Stakeholder' has arisen as stakeholder theory is open to many interpretations and applications ranging from business ethics and CSR to strategic management, corporate governance, and finance (Miles 2017). The author (Miles 2017) classified the stakeholder concept using a 16-category model based on an analysis of 885 definitions of stakeholder theory. Both influence-based (can affect) definitions and recipient-based (is affected by) definitions of stakeholder theory are discussed in the article by Miles (2017), and both genres are included as stakeholders in this current study. Stakeholder theory contends that businesses do not have solely to consider the interests of shareholders, but also account for other parties having a 'stake' in the decision-making of the company (Thijssens et al. 2015). Primary stakeholders in this context include employees, customers, competitors, investors, and suppliers, while the secondary stakeholder group includes governments, public interest groups, and the media (Matuleviciene and Stravinskiene 2015). Driven by social media and the Internet, public interest groups are becoming increasingly aware of sloppy global practices by corporations in (un)sustainable business activities. One such example is the Volkswagen (VW) scandal in 2015 whereby the giant car manufacturer said that it had understated the levels of carbon dioxide emissions in up to 800,000 cars sold in Europe (Wearden 2015). The challenge for the company now is dealing with the backlash from the public as it tries to reclaim trust and renew its dented Public Relations (PR) image. A wind turbine investment decision has consequences not just for the business investing, but also for the turbine suppliers and for the employees of that business. The turbine investment also has a bearing on the lives of the people living in the communities in which the company (and turbine) operates (Langer et al. 2018). Government bodies that may have assisted with the investment via grants or subsidies also have a stake in the venture. These stakeholder groups have a vested interest in the outcomes of this study. The unique closed-loop feedback framework which emanated from this research offers the opportunity to contribute in a positive way to group (corporate) decision-making regarding wind turbine investments made as part of a company's CSR initiatives. Demir et al. (2016) suggest that the reason for the secure link between stakeholder theory and CSR is that managers make decisions regarding activities within the business that affect a range of stakeholders and not solely the shareholders of the company; this agrees with the kernel of CSR. The study by Demir et al. (2016) analysed 176 sustainability reports of firms operating in Turkey published on the GRI website where trends in CSR reporting statistics were identified. Stakeholder theory origins may be linked to the spiritual sphere and the concept of the '*common good*' and the idea that all humans on this common earth are inextricably linked (Pope Francis 2015). In Pope Francis's encyclical on the environment, *Laudato Si'*, this idea is taken one step further where it is claimed that the natural environment and the human environment deteriorate together and that environmental degradation cannot be adequately combated unless we attend to causes related to human and social degradation (Pope Francis 2015, Chapter 1, V, 48, p. 29). The Pontiff claims that even though science and religion have distinctive and different approaches to understanding reality, it can be fruitful to both to enter intense dialogue in the move towards an integral ecology and the full development of humanity (Pope Francis 2015, Chapter 2, 62).

2.2.1.5 Social Obligation Perspective

Bowen (1953) stated that the doctrine of social responsibility is based on the idea that business should be conducted with a due concern regarding the effects of company operations upon the attainment of valued social goals (p. 8). Subsequently, Carroll (1979, 1991) defined social responsibility, stating the entire range of obligations that business has to society were primarily economic, legal, ethical, and discretionary categories of business performance. Mea and Sims (2019) present a conceptual framework, guided by Catholic Social Doctrine and Teaching (CSD/T), to improve business ethics with the ultimate purpose of serving human needs. The authors claim that business organisations provide a context for people to serve each other, to bring together capital and creative ingenuity in a social context to serve others. Businesses benefit in the broader social context when they are guided by moral norms.

2.2.2 Ambiguous/Subjective Nature of CSR/Sustainable Development

In the business community, it appears that the term 'Sustainable Development' is 'nebulous and contested' as stated by Sandelands and Hoffman (2008) with a general acceptance that the more familiar CSR term falls under the remit of Sustainable Development (Kealy 2016). Sustainable Development is defined by the Bruntland Commission as '*Development that meets the needs of the present without compromising the ability of future generations to meet their own needs*' (United Nations 2017). Research by Hansen et al. (2014) found that sustainability management and CSR have become tightly integrated. The authors claim that businesses have sensed the increased importance of environmental, social, and ethical issues. Both the financial and non-financial segments of business management are shown in Fig. 2.1. Of the three discrete parts—the three-P's, People, Profit, Planet—(Elkington 1997), financial reporting is the most established entity. The financial report gives periodic (quarterly, half-yearly, or annually) details of the financial activities and position of a business. The other two non-financial (human and environment) aspects of the reporting frameworks are less well established (Frias-Aceituno et al. 2013). Some of the Irish-based businesses who contribute to this study are aligned to the 'Origin Green' sustainability reporting framework from Bord Bia (2015). The Origin Green sustainability plan covers in detail six specific areas:

- Raw Material Sourcing,
- Energy Usage and Reduction,
- Water Conservation Using Rainwater Harvesting,
- Waste Product,
- Biodiversity,
- Social Sustainability.

Fig. 2.1 CSR/TBL
framework (Elkington 1997)

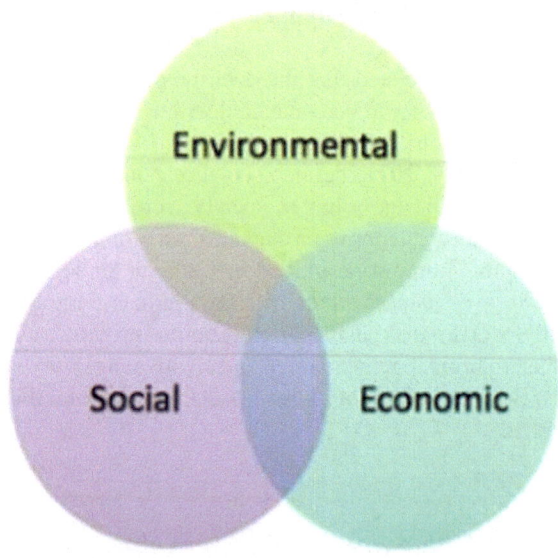

However, it is argued by Milne and Gray (2013) that sustainability reporting by itself, does not contribute to the sustaining of the earth's ecology. They claim that many organisations use the fact that they are reporting in the first instance, to carry on as usual without making any significant changes to their business practices, which paradoxically may lead to higher levels of *un*-sustainability. To improve the validity, or otherwise, of sustainability reports, Van Der Ploeg and Vanclay (2013) propose a sustainability reporting assessment 10-question checklist as a functional tool for use by stakeholders to evaluate the content of their sustainability reporting. Among the issues addressed by the questionnaire were (i) use of evidence to support claims and (ii) documented impacts of all stakeholders, including vulnerable groups and negatively affected groups. The authors use the internationally established Global Reporting Initiative (GRI) in its testing of the effectiveness of the checklist on an authentic but anonymous company.

While there have been many different theories, perspectives, and terminology in the CSR area since the 1950s, a specific connotation of CSR has not been unified (Saeidi et al. 2014). It has been claimed (Seaidi et al. 2014) that CSR and ethical issues must be incorporated into the curricula of business schools which would help to integrate CSR issues into corporate strategic planning and routine operational performance. Wang (2015) claims that CSR theory has remained controversial and ambiguous and has not yet fully matured. Different cultures also appear to have different ideas about the merits of CSR initiatives. For example, Buyaert (2012) claims that businesses in China are less concerned about a firm's activities related to CSR than their Western counterparts. This belief is reflected in the incoherent Chinese environmental laws and regulations that are enforced inconsistently across different regions of China (Buyaert 2012).

Attempts by Paredes-Gazquez et al. (2016) to aggregate indicators for the different CSR components of CSR into one measurement, found that the construction of a Composite Index (CI) was a complicated task; they claimed it might be impossible. A study of 556 Corporate Environmental Initiatives (CEI) announcements of Chinese firms over ten years between the years 2005 and 2014 (Lam et al. 2016) found that Chinese investors react negatively to such CEI announcements. This negativity may appear strange, but the market reaction in Western countries is quite different to market reactions in China, for example, Lyon et al. (2013) found that winning environmental awards in China has no effect and, in some cases, even harms shareholder value. Chinese investors are likely to believe that environmental initiatives and business objectives work against each other.

2.3 Stakeholder Responsibilities in the CSR Space

2.3.1 Business Practitioners

From the literature, it appears that some businesses have embraced the TBL philosophy and integrated the company CSR strategy within the company business strategy (Battaglia et al. 2016). In a study by Abro et al. (2016), the CSR values and practices in a leading Saudi Arabian oil company were analysed. This research case by Abro et al. in 2016 adopted Elkington's TBL model (Elkington 1997) to assess the economic, social, and environmental domains, as shown in Fig. 2.1. However, it should be noted that the robustness of this specific study by Abro et al. (2016) may be open to challenge, as the company in question is wholly owned by the Saudi Arabian government and works closely with universities in that country. Kurucz et al. (2017) claim that sustainability challenges are highly complex, with deep interdependencies between environmental, social, and economic factors.

In the economic sphere, the oil company in question (Abro et al. 2016) has remained very profitable by investing in best-in-class oil, gas, refining, and chemical facilities. It was found that these significant financial investments also provided direct economic benefits for Saudi Arabia. The company has also formed partnerships with other businesses to develop refineries in the area. In the social sphere, the study also found that this oil company had numerous initiatives in place to improve the local community. These included financing the region's infrastructures, providing better health care, building public schools that are safer, technologically smart and energy-efficient, also the building of a sports stadium, donating to orphanages, autism centres, and other institutions that serve disadvantaged Saudi citizens. In the environmental sphere, the company has adopted a twofold approach. It firstly reduced the environmental impact produced by its operations and secondly it conducted cutting-edge research. The environmental impact was reduced by complying with environmental standards and codes to protect air and water quality, public health, and safe waste management. The company also has a Carbon Dioxide (CO_2) reduction strategy in place that included promoting energy efficiency, protecting biodiversity, and targeting renewable energy. In terms of cutting-edge research, the company liaised closely with universities and made its Research & Development (R&D) facilities available to institutional researchers. It also made available financial support to researchers in this area. From the evidence presented in the case study by Abro et al. (2016) this Saudi Arabian oil company is keen to be perceived in good light with an excellent corporate image as the CSR division was situated in the Public Relations (PR) Operations Department. However, as previously mentioned, the findings of the study are to be accepted with caution as it does not state whether the research was independently funded, and it neglected to identify how the data was gathered. Oil (fossil fuels) is a negative contributory factor to global warming (Cullen 2013). Concurring with this study by Abro et al. (2016), research by Jung and Kim (2016) also found that firms tend to introduce CSR as merely window dressing. They utilised two different data sets to construct a comprehensive picture of the effects of CSR on employment relations in Korea. They found that as institutional pressure mounts for businesses to engage in explicit CSR behaviour, the subsequent CSR actions may not be an indicator that they are adopting CSR initiatives, but are designed to improve internal resources in response to the costs of undertaking the CSR activities in the first place. Managers look for improvement in efficiency from the workforce as a means of offsetting the initial CSR costs. They seek to improve a firm's reputation and secure its social legitimacy as a 'good neighbour' is in contrast with the actual reality of being a 'bad employer' by not looking after its employees. The study (Jung and Kim 2016) concluded by claiming that although companies are pressurised into behaving responsibly towards both the internal stakeholders and society, they are being hampered due to limited resources, and the competing demands of CSR make this balance difficult to achieve in practice (Jung and Kim 2016). The two different data sets used were (i) listed firms that were relatively large and had well-established CSR programmes in place and (ii) a representative sample of all firms with more than 30 employees in Korea. The reasons for using these two different data sets were to reduce bias, thereby achieving greater credibility and generalisability.

2.3.2   Social Responsibility of the Corporation

The seminal study by Bowen (1953) identified the social responsibility of the 'businessman' who leads the corporation. Bowen placed individual responsibility for attitudes, honesty, law observance, and business ethics paramount within the concept of 'stewardship'. Business leaders should be servants of society so that the idea of management merely in the interests of shareholders is not the sole end of their duty (Bowen 1953, p. 44). Bowen stressed that the corporation has a responsibility to be profitable, and there should be no attack on the profit motive or profits as such. He also discussed the responsibility of the state in helping businesses to use its power/position to bring about a more equitable society via its control over public policies such as taxation, tariffs, subsidies, and special tax concessions (Bowen

1953, p. 223). Bowen (1953, p. 3) argued that the decisions and action of busi-nessmen (and the businesses they direct) have a direct bearing on the quality and personality of people's lives. Constantine (2015) found that the quality of work–life balance correlates significantly and positively with workers' motivation to contribute to the Nigerian society. Xu et al. (2015) found that abusive supervisors can lead to subordinates becoming silent through emotional exhaustion. Bowen (1953) claims that businesses may determine significantly essential matters in society, such as the amount of employment and prosperity, the rate of economic progress and the distri-bution of income among various groups, and the organisation of industry and trade. Corporations also have responsibility for the morale of the labour force, the satisfac-tions obtained from work, the character of consumption, personal security, the rate of utilisation of natural resources, and even international relations. Bowen (1953) makes it clear that he believed that corporations have significant responsibilities and opportunities, to contribute to the well-being of society, a claim backed up more recently by Gutierrez and Vernis (2016). Part of this corporate responsibility is to carry out its business in compliance with the 'laws of the land'. Each corporation generally has compliance reporting structures in place (Gerard and Weber 2015). Gerard and Weber (2015) utilised data from eight compliance surveys over multiple years, to confirm trends and current issues regarding the state of compliance. They examined the effects of positioning the compliance function at various locations within an organisational structure. The mixed-method study examined the following core corporate responsibilities:

- Compliance with domestic regulation,
- Compliance training,
- Code of conduct (development and maintenance),
- Complaints and whistle-blower hotlines.

Gerard and Weber (2015) identified advantages and disadvantages in locating a corporate compliance function at different levels of the organisational structure. They also noted a perception that the regulatory environment is becoming more complex. They recommended that compliance functions should be better aligned strategically with CSR issues within the organisation (Weber and Gerard 2014) and that the impact of compliance should be measured. They claim that it is better to measure what compliance training achieves rather than just counting the number of compliance training sessions (Gerard and Weber 2015). While concurring, Kealy (2016) concluded that measuring sustainability impacts can indeed be an onerous task.

While it is acknowledged that group decision-making processes play a critical role in determining the performance of a business (Riccobono et al. 2016), one of the temptations within a consensus opinion is that group decisions may succumb to a groupthink mentality initially proposed by Janis (1972). Groupthink is defined by Janis (1972, p. 9) as 'a deterioration of mental efficiency, reality testing, and moral judgement that results from in-group pressures'. There is a danger that group dynamics can lead to concurrence seeking behaviour, and shared illusions (Janis 1972). More recently, Benabou (2013) claims that a groupthink mentality had a part to play in the Enron scandal and the 2008 global financial crisis. This phenomenon of groupthink may accentuate where there is a shortage of empirical data. Senior managers make many decisions, sometimes difficult ones involving complex and uncertain information (Kelman et al. 2017). Unclear information and a shortage of practical knowledge concerning wind turbine investment projects in Ireland were identified following a review of the literature. Although group decision-making is an iterative process in which selection and consensuses are interwoven (Choudhury et al. 2006), it is suggested that to insert some consensus control into the process is a desirable thing to do (Nikas et al. 2018). In the business community, employees traditionally work as team players and in groups (Costa et al. 2014). This team effect has the potential to overlook an individual's critical thinking, resulting in a desire not to 'rock the boat', as the individual may be seen as an outsider and encourages shared stereotypes of others (Sims and Sauser 2013). This culture of unquestioning deference may also have permeated through the host country, Ireland, and companies who take on renewable energy project may not thoroughly have investigated the potential benefits, financial, or otherwise of such ventures. It could be argued that the past number of years in Ireland has seen such a culture mushroom throughout many aspects of Irish society, business, political, media, church, social, with many commentators declaring that this relativism, accepting culture is destroying critical thinking, analytical skills with a reluctance to think outside the box or indeed to state an opinion outside the box (Sims and Sauser 2013).

2.3.3   Individual Responsibility

Businesses are managed as a collection of relationships between people (Jonczyk et al. 2016). It is people who drive and direct companies. This feature gives power to individuals to direct companies in the manner which they see fit. In a study by Osagie et al. (2016), the authors discuss the competencies that an individual must possess to support CSR implementation in a corporate context. The research methodology in this study included conducting interviews with 28 CSR directors and managers following which the authors identified eight distinct CSR-related competencies that the person should possess. One of the identified skills was to understand CSR-related systems while another desirable ability was to be able to balance personal ethical values *and* business objectives. The authors (Osagie et al. 2016) carried out a detailed literature review and initially identified 1229 scientific articles that would help to contribute to the authors' determination of the individual CSR-related competencies.

Gonzalez-Rodriguez et al. (2016) suggested that the different social, educational, and economic background of individuals directly influence consumer perception of CSR. In their research study using the survey methodology, 1200 Spanish social science students completed a questionnaire that included three well-differentiated types of questions. These were questions related to value priorities, inquiries related to attitudes towards a social initiative, and background variables. Rodriguez et al. (2016) suggested that universities should increase their awareness of the importance of integrating CSR and human values in the curricula of future business managers and social leaders. Frostenson (2016) discussed the virtue of humility and suggested that managers who possess this virtue can represent what the company is about in a truthful and logical representation. Frostenson (2016) claimed that humility is conducive to behaviours or attitudes among followers, which are beneficial to the company or leadership and can be a shield against public adulation. Steinmeier (2016) in a semi-structured interview qualitative piece of research, claims that sustainability managers in businesses face mounting pressure in their roles that can lead to what the author calls 'sustainability fraud' concerning their sustainability work. The fraud is referred to as deliberate actions taken by managers to deceive, con, swindle, or cheat investors or any other stakeholders. Steinmeier (2016) recruited three categories of interviewees for the study, namely, sustainability managers, sustainability experts, and forensic experts. Most of the interviewees to the survey believed that there is some manipulation in sustainability reporting, or greenwashing, and it is mainly done by companies limiting their reporting to positive aspects of the business. Sustainability managers face mounting pressure and have opportunities to manipulate due to an immature control environment. With the advancement of Multi-National Enterprises (MNE) and Mergers and Acquisitions (M&A), there is tension as to who has real control of the decision-making process within such organisations (Mehdi 2015). This global development has the potential to shift the responsibility away from the individual towards a group decision-making process. This has been a feature of large American corporations for a number of decades as stated by Bowen (1953, p. 84) who argues that with the separation of the functions of ownership (in the sense of legal title to equity capital) and control (in the sense of direction or management) within large corporations it can be tempting for the shareholders to confine their concern to the price of shares on the stock market and actual or prospective earnings. This focus has led to effective control of the large listed corporations to be passed into the hands of professional salaried managers. Bowen (Bowen 1953) discusses how this situation might bring different results and motivation to a corporation under owner/manager control. In a similar vein Buckley and Strange (Buckley and Strange 2015) in a review study, introduce the concept of the global factory and asks the following pertinent questions: (i) Who has control? (ii) Who captures the profits from global value chains?

2.3.4 State Responsibility

In recent years, globalisation has been eroding the regulatory power of the nation-state (Kohmar and Beschorner 2016). This erosion has led to a debate about 'global governance' which would incorporate nation-states, companies, consumers, and global civil society. The challenge is to try and get all these groups to interact and enforce rules in times of globalisation. Kolmar and Beschomer (2016) claim that up to now little traction has been generated in this area as the current status of responsibility that is assigned to either individuals (theories of morality) or a layer of institutions such as the state (theories of justice) or a corporation (theories of business ethics) is simply too limited. The authors claim that an alternative multilevel approach based on sound morals may be warranted which would locate responsibilities.

Buckley and Strange (2015) refer to the significant changes in the global location of economic activity as a 'global factory', driven by market liberalisation, financial deregulation and the integration of world financial markets, technology advances and the spread of preferential trade agreements. However, despite these significant changes, Anand and Segal (2008) argue that it is not definitive whether globalisation has increased or decreased global income inequalities. In their significant review piece considering Gross Domestic Product (GDP) and inward and outward Foreign Direct Investment (FDI) for 26 emerging economies which included 156 developing economies and 32 advanced economies, Buckley and Strange (2015) bring together insights from the literature on the governance of the global factory and its impact on economic development. They consider how global governance influences the distribution of income and query where the current responsibility for effective control of the globally dispersed economic activity lies. It is likely that this absence of global power could lead to unethical business practices. Nowhere has this been more prevalent than in today's global business activity. In Ireland, recent events have highlighted this phenomenon, a disagreement arose (September-2016) between the European Commission and the US technology firm, Apple. Apple was ordered to pay €13bn to Ireland in August 2016 after the European Union ruled that tax breaks it received from the Irish government between 1991 and 2015 amounted to unlawful state aid (The Guardian 2016). The ruling is likely to have a significant effect on major US companies investing in countries in Europe in the future, as they become concerned that a regulator can impose higher tax bills retroactively even though they were legal according to national laws. This so-called global factory scenario meant that in the case of Apple Inc., profits from iPhones or iPads sold in London, Paris, or Milan were subject only to Irish tax laws, even though Apple had minimal operations on the ground in Ireland, but the devices were 'booked' through a Cork Irish subsidiary, which meant that a low Irish corporation rate of tax applied. The European Commission ruled that Ireland gave Apple illegally favourable tax treatment, allowing Apple to pay an effective tax rate on European profits of 1% in 2003 and down to 0.005% in 2014 (Webb and Satariano 2016). The Chief Executive Officer (CEO) of Apple Tim Cook argued that they received guidance from the Irish tax authorities on how to comply correctly with Irish tax law and pay all the taxes that they owe in Ireland and in every country where they operate. It is clear from these recent events, and as identified by Langer et al. (2016), that transparent policy decision-making is critical for project acceptance by residents. This study by Langer et al. (2016) used data from both a literature review and expert interviews in the local federal state of Bavaria in Germany, data that was focused mainly on the renewable energy industry, but its robust approach would suggest that its findings could be applied to other industries in other countries. Schrempf-Stirling (2018) argues that the traditional roles of state and business have eroded with states losing power and business gaining power in a globalised world. The author argues that the future of

CSR lies in political CSR with new global governance forms which are organised by mainly non-state actors.

2.3.5 Government

A dictionary definition (Collins Dictionary 2016) informs us that a government is an executive body of public representatives tasked with exercising political authority over the actions and affairs of a political unit, as well as governing the performance of specified functions for this unit or body. One of the methods the government uses to guide the political unit (society) for the common good is to govern policy. Policy is a specific course of action that guides decisions to enhance the outcomes for persons in the government's jurisdiction, similar to strategy development in the private sector. Resources are put in place to implement the chosen policy. Different governments tend to have different priorities so therefore, may employ different policy initiatives. Policymaking is a practical process of transforming an idea into action. For the economic system to work effectively and efficiently, government can involve itself in the following issues:

- Market monopolies and competition laws (Howard 1950),
- Wealth distribution (Atkinson 2016),
- Economic growth (Mayer et al. 2016),
- Balance of payments (Lavoie 2015),
- Distribution of income (Atkinson 2016),
- Care of the environment (Wen et al. 2016).

The importance of effective government policy is the subject of a review paper by Shakeel et al. (2016). The authors claim that the power sector of Pakistan is in disarray, and they contribute this turmoil to the fact that the policies the government implemented did not work. Policymaking within government is influenced by many stakeholders. In a democracy, every citizen over 18 usually has a vote, thereby having a direct influence on which political party is elected into government. Influence can also be exerted on government policy decisions by groups who '*lobby*' the government. Dahan et al. (2015) claim that the government's place in the stakeholder conceptualisation of the business-society interface has in general, been very limited. This limited, light-touch, government intervention has been challenged in the wake of the 2008 global financial crisis and ongoing economic recession leading to calls for increased government involvement in the economy (Reich 2009). The study by Dahan et al. (2015) analysed typical governmental activities/behaviours and relationships with firms stakeholders and subsequently described governments as serving four distinct roles relative to business and its related stakeholders. These included, namely, 'framework', 'business partner', 'interfering', and 'advocate'. This study suggests that the political stability of a country, therefore, can have a direct influence on business and CSR activity.

In a study employing an ethnographic approach, Ptak (2019) explored the uneven distribution of benefits resulting from government-led programmes to utilise hydropower in rural electrification policies in China. The unequal distribution of benefits is attributed to the fact that China is such a diverse country, geographically, culturally, and biophysically. The ethnographic study included household surveys, interviews with local residents, and focus groups. In conclusion, Ptak (2019) argues that government policy did favour some geographical areas over others, spatial privileging and patterns of uneven development did exist. One method of overcoming this uneven development is to carry out more human-centred ethnographic studies.

2.3.5.1 Free-Market Economy

A free-market economy is an economy in which the allocation for resources is determined only by the supply and demand for them (Garen 2010). In theory, the government plays no role in organising the flow of resources to the production of various goods and services. In practice, however, there is no such thing as a free-market economy. Every government places some restrictions on the ownership and exchange of commodities; this is called a mixed economy where privately owned businesses and government both play important roles. It is generally thought that some services are better performed by public bodies rather than private corporations, for example, the health service, education system, justice system, road infrastructure, and defence are mainly provided by a public service. Reduced governmental involvement in the free market was championed by Margaret Thatcher in the UK and Ronald Reagan in the USA in the late 1970s and early 1980s. A study by Ullah (2016) assesses the expected benefits of the free-market economy on manufacturing workers in developing countries (Bangladesh). However, the author found that the economic benefits and increased social status did not materialise. The free-market capitalism, in this case, resulted in income (in)equalities and decline of workers' emancipation. The author claims that greed has tainted the free-market idealism, especially in developing countries, and that the geopolitical system supports this unequal capitalist culture. This viewpoint is supported by results of a study by Foellmi and Oechslin (2010) who found that free trade significantly widens income differences in less-developed countries. They claim that wealthy entrepreneurs become richer while more impoverished business people are worse off because of trade liberalisation.

2.3.5.2 Governmental Policy in the Energy Sector

It is clear from the literature that government policy in the field of renewable energy has a significant influence on its development (Chapman et al. 2016; Cullen 2013). Chapmam et al. (2016) used Qualitative Content Analysis (QCA) on energy policy reports, academic papers, and government publications for 8 OECD nations, to reveal common themes and patterns within the data. For each of the participant countries, the energy policy goals are (i) an increase in renewable energy-based electricity generation and (ii) a reduction in Green-House-Gas (GHG) emissions. The tools in place to achieve the energy policy goals are in all cases, economically based and include the following:

- Feed-In-Tariffs,
- Incentive payments,
- Tax or Depreciation concessions,
- Monetary non-compliance penalties (Chapman et al. 2016).

The government can play a role in providing equity in the market through taxation, regulation, and subsidies (Dragomir et al. 2016). Concerning caring for the environment, national governments can, through the Energy or Environmental departments, offer support by providing subsidies/grants/loans, and financing programmes for renewable energy systems. These policies are intended to improve energy security, energy affordability, create jobs, and minimise the negative effect that the burning of fossil fuels in traditional electricity generation plants has on the environment. Highlighting this from a Polish perspective, Iglinski et al. (2016) discussed a law enacted by the Polish government in 2014 in the area of renewable energy sources. This law stipulated new regulations and conditions which would guide support mechanisms, thereby encouraging the generation of electrical energy from renewable sources (Iglinski et al. 2016). The law states that a seller is obliged to purchase electric power from newly built renewable energy sources. This policy is funded by an extra fee on an energy company's tariff for each electricity customer.

In Ireland, the government is committed to reaching its binding target of 16% overall renewable energy, of which 40% penetration is targeted in renewable energy for electricity consumption (Sustainable Energy Authority of Ireland 2012). As the development of wind power in Ireland is at an early stage, the Irish government plans to assist the industry for some time until it gains competitiveness and can stand alone economically. This assistance comes in the form of subsidies, funded through the Public Service Obligation (PSO) framework, and given to businesses so that they can provide electricity generated from renewable sources, such as wind. The Single Electricity Market Operator (SEMO) is the operator for the all-Irish electricity market. Licensed electricity generators sell their generated electricity to an authorised supplier who sells it onto a central pool system for a Single Market Price (SMP). Every electricity generator is given a daily schedule of how much power they can dispatch to the grid. As the supply and demand for electricity fluctuate during the day, perhaps due to changing load conditions or changing wind conditions, the SEMO can order the electricity generators to increase or decrease how much they are supplying at a moment's notice (Single Energy Market Operator 2016).

The government authority overseeing the rules governing the electricity market in the Republic of Ireland is the Commission for Energy Regulation (CER) (Commission for Energy Regulation 2016). The CER must require that the system operator gives priority dispatch to generating stations using renewable, sustainable, or alternative energy sources. The Renewable Energy Feed-in Tariff (REFIT) is a scheme designed to support the wind industry by providing price certainty for units of electricity generated from renewable sources, such as wind. This scheme offers a guaranteed price for a unit of energy (kWh) plus an additional payment of 15% on top (Commission for Energy Regulation 2016). In an empirical piece of research by Midttun et al. (2015), they warn of the conflicts that can arise when advanced welfare states introduce CSR into public policy. Their concerns centred on the potential danger of leaving key public welfare issues to the discretion of private businesses. The study (Midttun et al. 2015) was conducted using interview data from 55 officials from government ministries, Non-Governmental Organisations (NGO's), labour unions, and employer associations in Denmark, Finland, Norway, and Sweden. Their research concluded by highlighting the fact that tension does indeed exist between introducing CSR into public policy and the traditional method of controlling corporate conduct using negotiated agreements and strong regulation. There is a harmony of goals but conflict in means of achieving those goals (Midttun et al. 2015). In terms of governmental policy in China, an empirical study using data from 30 provinces and municipalities between the years 2004 and 2013 by He et al. (2016) concluded by claiming that policymakers should focus on strengthening governmental regulation to drive forward an affirmative corporate responsibility and also to encourage companies to reduce the number of pollutants emitted into the natural environment as a result of their business operations. This study (He et al. 2016) may be limited by the fact that there are different cultural expectations in different nation-states. These diverse cultural differences are highlighted in a study by Karaibrahimoglu and Cangarli (2016) who used regression analysis on data regarding cultural values and ethical behaviour of firms in 54 countries between the years 2007 and 2012. Data for the research (Karaibrahimoglu and Cangarli 2016) was gathered from the World Competitiveness Index (WCI) database through the World Economic Forum's Executive Opinion Survey which gives both a qualitative and quantitative portrait of countries' economic and business environment.

It appears that there are two main limitations to governments acting as the ultimate public policy decision-makers, or social referee (Dahan et al. 2015). Firstly, they claim that there is a lessening of national sovereignty owing to international trade treaties which includes corporate disputes being settled by independent bodies and not local government. This means that differences between investors and governments would not be settled in the country where the disagreement arose, but by independent arbitrators (Coy et al. 2014). Secondly, there appears to be a social movement evolving contrary to an over-involved national government based on what is perceived as an excessive interference of government in the economy, the so-called 'Nanny-State' (Dahan et al. 2015).

2.3.6 Local Communities

The acceptance by local communities of wind energy technologies is vital to the success of such projects (Langer et al. 2016). In a qualitative study, using the interview research methodology conducted with a range of stakeholders in the German region of Bavaria, Langer et al. (2016) identified the factors that influenced the acceptance of wind energy in that region. Factors that emerged from this study included (i) the distance of wind turbines to the place of residence, (ii) the visual appearance of wind turbines, (iii) involvement of citizens and societal actors in the project mainly in the participation of profit or rental income, and (iv) the perception that the political process towards sustainable energy development is shaped by trust, transparency, and the perceived fairness of the development process. Langer et al. (2016) found that some of the local community not directly involved in the wind energy project can feel envious if they suspect that the wind turbine owners are getting a good profit on the investment while they are left out. Supporting these findings and within an Irish and Scottish context, Warren et al. (2005) carried out two case studies using face-to-face questionnaire surveys conducted at residents' homes which explored the public perceptions of onshore windfarms in those areas. A vast majority of the negative responses within the study's results were attributed to the wind turbine being unattractive, the aesthetic nature of the turbines being the most potent single influence in public attitudes (Warren et al. 2005). There were some positive responses to the acceptance of the turbines, particularly in Ireland, with the NIMBY (Not-In-My-Back-Yard) effect diminished over time. The worry by residents that noise pollution could be a factor was not realised in this case. It is claimed that some wind turbine developers and local communities use their political power by lobbying government personnel to guide activity in the alternative energy area (Lock and Seele 2016). Research by Wolsink and Breukers (2010) found that a more collaborative perspective on decision-making about wind power increased the chance of successful outcomes in the venture. Host communities have certain expectations regarding involvement and revenues.

Lakhanpal (2019) employed ethnographic methods to analyse the case of local opposition to a 113-MW wind power project in India. The research was conducted using ethnographic fieldwork which included semi-structured interviews at the field site in three villages and one city. The study found that contestations to renewable energy projects in developing countries centre around issues such as access to natural resources and affect local communities who have nature-based livelihoods. This issue is a different contestation from the long-observed NIMBY in developed countries, usually attributed to aesthetics. One of the ways to overcome the contestations is to give local communities greater bargaining powers along with mandatory social and environmental impact assessments (Lakhanpal 2019).

2.3.7   Media

The news media has a significant effect on how businesses conduct their activities (Bednar 2012). They occupy a unique position as information intermediaries between business and society. Bednar (2012) examined the agenda-setting theory to assess how the news media influences public opinion. It appears that the media decide *what* the public should think about and *how* the public should think about specific issues. Jia et al. (2016) suggest that news media is profit-driven, and because of this, they naturally report news that attracts public attention. Individuals tend to pay more attention to negative information, as they consider it to be more informative. In the environmental sphere, the news media reported extensively on (i) the nuclear catastrophe in Fukushima in Japan in 2011 where a tsunami triggered a nuclear meltdown and the release of radioactive material into the atmosphere, (ii) the BP oil spill in 2010 when an explosion on an oil rig resulted in oil gushing into the sea for approximately 87 days, and (iii) the Volkswagen CO_2 emissions scandal in 2015 when the car manufacturer was found to have software installed in the engines that could detect when they were being tested and subsequently report improved carbon dioxide emission results (BBC 2016). Jia et al. (2016) claim that negative news coverage of business after irresponsible business practices generates an atmosphere of negativity in the general public towards the company. Managers in these companies subsequently commit to correcting these practices to ensure corporate survival and preserve or reassert their reputation. Jia et al. (2016) studied news media reports from more than 600 newspaper sources regarding listed Chinese firms between the years 2004 and 2012 and concluded that news media could play an important role, along with the government, in helping to mitigate irresponsible environmental actions by businesses. A limitation of the study by Jia et al. (2016) is that the investigation is based in China where the Chinese government sensors certain news topics and also it examines the specific environmental problem of corporate pollution. However, questions may be asked if there is there any such thing as independent news media without an agenda to pursue? Further study was recommended by Jia et al. (2016) to investigate the financial and political connections of news media, as it was suggested that influential media/business relationships might shield firms from negative coverage when undesirable behaviour occurs. They also indicate a need for further study in the role of social media in disciplining irresponsible corporate practices in the current era of global connectivity.

2.4 Empirical Measurements of CSR Initiatives—Sustainability Reporting

There are increasing demands from many stakeholders to provide transparent disclosure on economic, social, and environmental performance (Rasche and Esser 2006), and this disclosure is typically implemented in the form of sustainability reporting (Kealy 2019). However, Herremans et al. (2016) suggest that how reporting is explicitly used to engage stakeholders, is understudied. The authors investigate how companies address different dependencies on stakeholders for economical, natural environment, and social resources and thus engage stakeholders accordingly. The diversity in sustainability disclosure is explained by finding out how the report is used to engage different stakeholders. Motivations for engaging stakeholders include the securing of capital to conventionally fund the business (also to be seen as worthy of investment in socially aware markets), to meet political and social expectations, and also for organisational learning. The transparent sharing of knowledge by the publication of the sustainability report can be beneficial for decision-making within the business (Herremans et al. 2016).

2.4.1 Financial (Economic) Component

Annual corporation financial reporting and auditing that describes company performance outcomes is a well-established corporate activity. Standard financial reports cover the basic financial statements, namely, balance sheet, statement of income, statement of retained earnings or changes in stockholder's equity, and statement of cash flows (Abernathy et al. 2019). It is important that financial results are reported in an accurate and honest way. Audit committee members are entrusted with the responsibility of protecting shareholder interests and overseeing the external audit process. Challenges arise when a company may have to present negative financial statement events and audit committee members may have to choose between protecting their self-interests, as opposed to investor's interests (Kang 2019). An emerging practice that has developed is for businesses to produce annual reports that integrate financial and non-financial indicators (Maroun 2019). The non-financial aspects have been traditionally seen as sustainability/CSR disclosures. There is an ongoing debate as to the financial (Economic, Fig. 2.1) benefits of embracing sustainability/CSR initiatives. Madorran and Garcia (2016) investigated Spanish firms on the IBEX35 Stock Market index between 2003 and 2010 and concluded that there is a neutral relationship between CSR and financial performance. The authors identify measurement problems, particularly that of the CSR variable, as one of the possible causes for this finding. Another study by Nollet et al. (2016) also reported mixed results when evaluating the relationship between sustainability/CSR and financial performance.

2.4.2 Human (Social) Component

The social component of sustainability refers to human ecology, those aspects of a business that contribute to society (Mea and Sims 2019). In this context, society includes customers, investors, employees, and the communities in which a business resides. The research by Mea and Sims (2019) is based on the belief that every human being has a transcendent value that resides within his or her essence, an essential aspect of what makes them a person. The ultimate purpose of business organisations, as dynamic networks of people in society, should be to serve human needs. In terms of the human (Social, Fig. 2.1) benefits to CSR activities within businesses, the findings of a study by Dumitrescu and Simionescu (Dumitrescu and Simionescu 2015) suggest that there may be a positive influence and significant relationship between CSR activities and the attitude of company employees. This positive influence is reflected in company performance. In this study of listed companies in Romania conducted over four years, the authors carried out regression analysis to probe the link between the two variables. Earlier work by Venter et al. (Venter et al. 2014) also concluded that employee satisfaction is significantly enhanced by a Small–Medium Enterprise (SME) engaging in CSR activities. Venter et al. (2014) conducted a quantitative study utilising structural equation modelling on 383 respondents to a self-administered questionnaire which was distributed to SME's in Uganda. These results suggest that satisfied employees help to enhance the competitiveness of the SME in the long run. Research by Searcy et al. (2016) analysed the relationship between the employees' work environment and its influence on workers' health and organisational performance. The study was based on indicator disclosures in company CSR reports. However, the study found that different companies used a wide range of work environment indicators and concluded by claiming that there is a need for greater standardisation in work environment reporting as part of the overall CSR reporting. Indeed, Staniskiene and Stankeviciute (2018) argue that while the Environmental and Economic dimensions of CSR (Fig. 2.1) could be more easily evaluated by clearly expressed quantitative indicators, the Social sustainability measurement requires a balance between quantitative and qualitative indicators. The link between human rights and business is discussed by Buhmann et al. (2019) where the authors connect the Sustainable Development Goals (SDGs) with business and human rights and political CSR theory. They found that companies can benefit from their (human) resources by drawing on insights gained by considering human rights. This focus allows the company to develop appropriate interventions to address local needs. The business interventions assist in fulfilling human rights issues and advance the business moral legitimacy by contributing to sustainable development goals.

2.4.3 Environmental Component

Publications in the environmental component of CSR have dealt with issues such as water security (Whaley and Weatherhead 2016), pollution (Galupa et al. 2014), deforestation (Tsurumi and Managi 2014), waste disposal (Ghoseiri and Lessan 2014), loss of biodiversity (Santos et al. 2015), and specifically in this review the phenomenon of climate change and global warming (Anderson et al. 2016). This issue of climate change currently exercises the most discourse and debate in current times (Carrington 2016). The decision to implement wind power projects to mitigate climate change and global warming and safeguard scarce resources is considered to be part of the environmental component of CSR (Wolsink and Breukers 2010). The burning of fossil fuels in traditional electricity power generation plants produces pollutants such as carbon dioxide (CO_2), nitrogen oxides (NO_x), and sulphur dioxide (SO_2) and is thought to be contributing to the well documented undesirable increase in global temperatures. SEAI (2017) calculated that 33% of primary energy use in 2016 was energy used for electricity generation (the remainder being 32% for Thermal and 34% for Transport). The 2016 value for gross generation of electricity increased by 2% to 26 TWh with a 7% increase in the fuel inputs. The primary energy input was 4812 ktoe (SEAI 2017, p. 20) of which wind contributed 529 ktoe and Natural Gas contributed 2334 ktoe. Concerning better electrical energy usage, it is generally accepted that there are two ways to influence this process. *Demand-Side Management* (DSM); use less energy in the factory to do the same job, and *Supply-Side Management* (SSM); seek to implement alternative natural energy sources. While the Irish government is keen to promote both initiatives to decarbonise its energy system, currently there appears to be more activity in the planning and building of alternative energy sources like wind turbines, with a major push for wind penetration to increase as part of these incentives (Table 2.2). This wind turbine activity is not without opposition, as some people have begun to question the benefits of wind energy. Some also argue that turbines cause damage to our environment and have the potential to divide local communities forced to live in industrial landscapes (Warren et al. 2005). It seems that despite the renewed interest in wind turbine projects by many stakeholders there appears to be a dearth of empirical research conducted into the outcome evaluation for initiatives in the environmental (Fig. 2.1) component of the overall CSR model. Some authors (Karassin and Bar-Haim 2016) suggest that one of the reasons for this may be that there is no single standard measurement system that accurately represents and allows comparison of environmental performance across different industries. Their study (Karassin and Bar-Haim 2016) involving regression analysis was conducted in 2013–2014 and involved 11 medium and large Israeli industrial facilities. Fifty-four top managers were interviewed, and 412 workers responded to a questionnaire as part of the data collection process. They claimed that the filling out of the sustainability reports which is carried out on a self-reporting basis has the potential to impose a limitation on data availability (that which is measured), transparency (that which is released or reported), and quality (accuracy and truthfulness of the data reported). Karassin and Bar-Haim (2016) claim that these limitations are most evident in countries where compliance databases are sparse. The development of a standard measurement system may require *collaborative* research involving engineers, managers, and economists. Despite the growing institutional, human, financial, and scientific resources made available nationally and globally in the fight against climate change, such as the €280,000 capital investment by the SME detailed in Chap. 6, GHG emission continues to rise at an unacceptable rate (IPCC 2013). Androde and Puppim de Oliveira (2015) suggest several factors that may explain the failure in the process of reducing emissions, one of which is the lack of implementation and effectiveness of the global climate and energy governance regimes. Governance of the climate change initiatives is generally of a non-regulatory manner, and it is left up to each corporation (and each Nation) to report voluntarily stating their non-financial aspects of its operations. Prado-Lorenzo and Garcia-Sanchez (2010) suggest that as the Triple-Bottom-Line (TBL) reporting framework whereby companies show accountability for three aspects of their activities (People, Profit, and Planet) has a voluntary nature, it could be targeted at other business objectives. The authors claim that, as the production of these TBL reports creates high costs, it may be logical to expect the companies to search for positive effects, other than just concealing their less appropriate practices in the field of sustainability. The effectiveness of the energy governance regimes in an Irish context is discussed in this book.

2.4.3.1 Climate Change

The increase in the average temperature of the Earth's climate system is termed global warming. The dominant neoclassical business growth model has put pressure on our vast, but finite, natural resources (Catalin and Nicoleta 2011). Traditionally, most of the electrical energy purchased by businesses was generated using fossil-fuel-(oil, coal, or gas) driven generators but it is now widely accepted that the GHG emissions (such as CO_2), due to the burning of fossil fuels, contribute to the phenomenon of global warming. It is therefore incumbent on business, from a corporate citizenship point-of-view and a Public Relations (PR) point-of-view that corporations endeavour to be responsible global citizens by embracing technologies that can reduce their dependence on fossil-fuel technology (Pinkse and Kolk 2010). The area of GHG emissions is a major political issue gaining attention on a global platform (COP 2015). Boiral et al. (2012) claim that social pressure to reduce GHG emissions is one of the main determinants or businesses commitment to climate change. Faced with increasing demand from stakeholders, companies feel the need to disclose information on climate performance and GHG emissions to legitimise their industrial activities (Hrasky 2012). A standard method of revealing information on climate change is through sustainability reporting (Perego and Kolk 2012). Sustainability reporting gives information on the non-financial aspects of the business operations, including the effect that their activities are having on the natural environment. By seeking alternative, renewable, energy sources to generate electrical power required in the business and therefore offsetting the amount of fossil-fuel-driven CO_2 emissions, it is anticipated that the effect is to slow down the phenomenon of global warming. Research by Besio and Pronzini (2014) claims that corporations have a moral obligation in how they respond to societal demands to take responsibility for climate change. They argue that in some occurrences, morality in business becomes a mere façade while in others, it serves as a decision-making criterion and profoundly influences the core activities in firms. Brusseau et al. (2013) also highlight concerns regarding the authenticity of some company's adoption of green energy strategies. They argue that many businesses are 'greenwashing', a process of appearing to embrace social and environmental stewardship within the business model, while the real intention is to convert that branded concern into money on the *Profit* bottom line.

2.4.3.2 Alternative Energy Systems (Wind Turbines)

This book mainly focuses on one specific aspect of SD/CSR, namely the environment, and more concisely one element of the environment, namely, climate change and how alternative energy systems can help in the battle against climate change by reducing harmful GHG emissions. Alternative energy systems include solar Photo Voltaic (PV) panels, Hydro-electric, Biomass, and Wind turbines. This study focuses on wind turbine installations. The wind turbine strategic investment by businesses is in response to Irelands' push to comply with the EU Directive 2009/28/EC, which mandates the levels of renewable energy use within each member state. The Directive sets a target for Ireland of 16% share of energy from renewable sources in gross final consumption of energy in 2020. Failure to reach this binding EU target leads to substantial fines being imposed on the Irish government. A breakdown of this EU overall target of 16% shows that 40% of electricity generation is due to renewable sources and this currently stands at 22.7% (Sustainable Energy Authority of Ireland 2015). Irish government energy policy has prompted recent intense activity in wind turbine proposals and installations in the hope that wind can be a significant contributing factor in reaching that 40% target. It is expected that every unit of energy generated by wind turbines 'offsets' pollution that would have been emitted by a conventional fossil-fuel generator (Cullen 2013). Wind energy turbines do not generally operate as stand-alone sources of electrical energy because of the variability and intermittency of the power source (wind); they require back-up generation such as gas generators to allow for the uncertainty and variability (Puga 2010). The power output of a wind turbine depends upon the air density and wind speed at hub height (Diaz et al. 2018). The stochastic nature of wind is exacerbated by the fact that wind power generation is proportional to the *cube* of wind speed (Kaffine et al. 2013). Therefore, if the wind speed halves, there is an eightfold decrease in power output generation. The wind turbine power output varies rapidly, even when several sites are connected (Katzenstein and Apt 2009). The back-up (embedded) generators operate more efficiently when working *steadily* near-maximum capacity; operating at partial capacity may increase emission rates. Emissions rates can also change during ramping (Cullen 2013). Periods, when a generator is ramping up, have higher than average emission rates and a lower emission rate when the generator is ramping down although the effect is not necessarily symmetric. Much of the literature utilises modelled wind turbine data to predict and estimate the CO_2 emission savings as a result of embracing wind turbine technology (Katzenstein and Apt 2009; Kose et al. 2014; Kaffine et al. 2013). There is a dearth of published literature demonstrating the actual *measured* benefits of embracing wind turbines as alternative energy systems. There is also a lack of research describing the quality of the power output from wind turbines. Such documentation would indicate the degree to which the power output is varying. The variations, ramping, and uncertainty in the output of wind farms present challenges to the national electricity system operator. The system operator must decide ahead of time what generators should be online to meet the load demand efficiently. This decision is based on forecasts of wind output and electricity demand, which may be unpredictable. The variations are typically managed by providing fossil-fuel generators as *spinning reserve* in waiting mode, ready to take over when the wind turbine output drops. Some fossil-fuel turbines need to be *spinning* as they can take several hours to warm up and come online. Therefore, the additional cycling and ramping and reductions in online capacity factors due to wind generation reduce the efficiency and effectiveness of individual fossil-fuel generators (SEAI 2012). The Sustainable Energy Authority of Ireland (SEAI 2012) stated the following points:

- With renewable electricity on the system, fossil-fuel generators spend less time generating for each time they start (additional cycling),
- Displacement by renewable electricity generation reduces the average output from fossil-fuel generators, indicated by a reduction in the online capacity factor of Combined Cycle Gas Turbine (CCGT) and coal-fired generators,
- Individual fossil-fuel generators run in less efficient modes with renewable electricity on the system, showing a 7% increase in the CO_2 emissions intensity for such generators.

In recent years, many SME's have installed on-site embedded/autoproducing wind turbines as part of the government energy policy to encourage businesses to supply some, or all, of their electrical energy requirements. A second significant advantage of embracing indigenous renewable energy sources is to help in reducing the country's' dependence on imported fossil fuel. A November 2015 report 'Energy in Ireland 1990 – 2014' commissioned by the Sustainable Energy Authority of Ireland (SEAI) claims that while there was a modest reduction in overall energy use by 0.5% in 2014, Ireland still had an import dependency of 85% in 2014, costing €5.7 billion. According to Eurostat, the European Union's official statistics body, only Malta, Luxembourg, and Cyprus fared worse in terms of imported energy dependency in 2014. The UK had an imported energy dependency of 45.5%, below the European average of 53.4%. The final consumption of electricity in 2014 was almost static in the previous year at 24.14 TWh's with a 0.7% reduction in the fuel inputs. The final electricity consumption figures were sourced from data provided by the CER. As part of its role, the CER jointly regulates the all-Ireland wholesale Single Electricity Market (SEM) with the utility regulator in Belfast. The data is collected from the retail market reports published by the seven current active suppliers in the electricity retail business and domestic markets. A 6-year summary of final electricity consumption was

- 24.14 TWh (2014),
- 24.2 TWh (2013),
- 24.2 TWh (2012),
- 24.9 TWh (2011),
- 25.4 TWh (2010),
- 25.3 TWh (2009).

2.5 Key Stakeholders in the Wind Industry, Ireland

Three of the leading independent developers and operators of wind energy projects in Ireland are Gaelectric, Mainstream Renewable Power, and Element Power. These companies also operate in the global renewable energy market. Gaelectric was founded in 2004 and currently, in September 2016 have 174-MW of wind generation connected to the grid. They are developing energy storage projects in Northern Ireland, the UK, and the European mainland (Gaelectric 2016). Mainstream Renewable Power was founded in 2008 by Eddie O'Connor who sold his previous company Airtricity to SSE and E.ON for €1.8 billion. Mainstream has offices in eight countries, across five continents and currently employs 140 staff (Mainstream Renewable Power 2016). Element Power was established in 2008 and had operations in eight countries. They develop, acquire, build, own, and operate a portfolio of onshore wind and solar power generation facilities worldwide (Element Power 2016).

Installed Capacity of Wind Generators in Ireland: The share of electricity generated by wind in 2013 was 16.4%. This value is calculated according to EU Directive 2009/28/EC and includes normalisation rules for wind. The EU Directive states that electricity from wind (and hydro) needs to be normalised to smooth out the effects of annual variations. Normalised generation is calculated using the weighted average load factor over the last five years for wind. In 2014 wind generation accounted for just 18.2% of electricity generated despite the increase in the number of wind turbine installations. The installed capacity of wind generation reached 2.211-GW by the end of 2014 and also there was 0.224-GW of wind generation introduced in 2015 (approximately 2.4-GW of wind energy generation capacity at the end of 2015) and a further 1.027-GW contracted by the end of 2016. The overall share of fossil fuels used in electricity generation was 80.8% in 2014, down modestly from 82.6% in 2013. The total power output of connected fossil-fuel-driven generators is approximately 6.23-GW of electrical power (Eirgrid 2015). Taking these figures into account, it is not unreasonable to expect more substantial savings and benefits, considering the increased Irish policy incentives for wind technology currently being experienced in Ireland. A lack of empirical data on the ramping effect of short-duration wind variation may contribute to the conflict between the *estimated* benefits of wind turbine generated power and *actual* benefits. A predicted financial appraisal by Kose et al. (2014) of a 6-MW wind farm claimed to have a Payback Period (PP) of 6.44 years, but this value was calculated using modelled data and may, or may not, be accurate. Empirical research by Kealy (2014) on a 10-kW three-phase wind turbine found a PP of 23 years using real data read on-site from the turbine output indicator. This *actual* measured data calculation contrasts with many *predicted* calculations, so caution needs to be applied when SME's are embracing wind turbines for CSR purposes. The owner of the wind turbine did not see any significant reduction in the number of imported energy units, kWh's after the turbine was installed. Another empirical piece of research by Kealy et al. (2015) investigated the output of a 3.5-MW wind farm in Ireland and found satisfactory outcomes in terms of the capacity factor and the number of kWh units produced annually. However, the annual kWh energy output from the wind farm, 9,808,318 kWh's, were metered digitally and fed directly into the National Grid (38 kV Sub-station) and there was no further independent research to find out if this amount displaced the number of kWh units from traditional fossil-fuel sources. This lack of empirical research is worrying and might be stifling intellectual debate on the issue. In the national, Irish, narrative, Waters (2015) in an opinion piece claims that there appears to be a '*lack of intellectualism*' and a '*group-think*' approach to many aspects of Irish society. The groupthink claim is one of the issues discussed in an Irish-based (and Denmark) study by Heaslip et al. (2016) who determine the enablers and barriers to implementing effective sectors sustainable energy communities. While influential organisations promote debate, plainly more detailed, independent, research needs to be carried instead of calls for 'consensus-seeking' conversations and listing corporations who have spent millions of Euro investing in wind turbine projects (Melia 2015) without an independent appraisal of these projects. Independent research is needed to make claims about turbine installations, and this research can contribute to that debate. This research investigates the effectiveness of a business wind turbine investment decision in contributing positively to the *Environment* part of the SME CSR activities. A wind turbine embedded electrical generator is expected to reduce GHG emissions by offsetting electrical energy units usually supplied by fossil fuel burning generators. It must be remembered, however, that wind power output is variable and when business owners/investors are making decisions of the appropriate wind turbine size and technology, it is useful to have information regarding estimates about resulting wind turbine variability (Boutsika and Santoso 2013). This book aims to inform interested parties in this regard.

2.5.1 Wind Industry—Global

While the first wind turbine for electricity production was built by Blyth in Glasgow in 1887 (Jones and Bouamane 2011), it is only in recent decades that a sizeable number of articles concerning wind turbines began to be frequently published. Although they were not the first country to use wind power to generate electricity, it was in Denmark that the wind industry began to fully develop as a business venture. In 1959 a Gedster wind turbine designed by a Dutch engineer named Johannes Juul began operating. The 200-kW turbine ran for ten years as the largest turbine in the world until it was shut down in 1967 (Jones and Bouamane 2011). The Danish wind energy industry was further intensified when agricultural equipment manufacturers such as Vestas and Nordtank diversified into wind turbine manufacturing in the 1970s and 1980s. These companies knew how to build heavy machinery for a rural market and used these competencies to good effect in the design, manufacture, construction, and maintenance of heavy industrial wind turbines (Jones and Bouamane 2011). While the global wind industry is smaller than the global conventional power generation technologies, it has received major interest in the past couple of decades (Xu et al. 2016). The global wind power industry had record-breaking years in 2014 and 2015 with a 22% annual growth rate in 2015 while the country with the most power installed during the year was China with 30.8-GW of new installations (Global Wind Energy Council 2016). Other countries to have a steady wind growth rate in 2015 were Canada, Brazil, and Mexico.

In a significant research piece from a European perspective, Gonzalez and Lacal-Arantegui (2016) present an overview of the regulatory framework for wind energy in the European Union (EU) member states. Firstly, this review paper presents the main aspects of each EU member states' National Renewable Energy Action Plans (NREAP), while secondly concluding with an analysis of actual developments of wind energy activity in each member state. One of the main aspects discussed in this piece was the support schemes for wind energy promotion. The support schemes discussed included (i) Feed-In-Tariffs; (ii) Feed-In-Premiums; (iii) Tenders; (iv) Quota obligations; (v) Tax incentives or exemptions; (vi) Investment grants; and (vii) Financing incentives. Commenting on grid connection, Gonzalez and Lacal-Arantegui (2016) highlight other issues such as cost allocation, priority use of the grid, and potential barriers for wind energy deployment. They concluded that there is a strong link between a favourable regulatory framework and actual deployment of wind energy. Some EU member states have a strong commitment to supporting wind energy while other member states have not provided enough support to stimulate the desired level of investment.

2.5.2 Wind Turbine Manufacturers

Recent peer-reviewed published research on wind energy, for the most part, does not identify the specific turbine manufacturers in their publications because of ethical issues surrounding anonymity. However, this researcher feels that it is essential to explicitly name the leading global wind turbine manufacturers as they are major stakeholders in this CSR field. The information for the top global wind energy businesses is taken from what is claimed to be the leading independent news magazine reporting solely on wind energy matters entitled Wind Power Monthly (Wind Power Monthly 2016). The magazine has been reporting nonstop since 1985. The top ten wind turbine manufacturers account for almost 270 GW of global installed capacity and much of the recent technological progress (Wind Power Monthly 2016). The leading players are shown in Table 2.1.

2.5.3 Wind Turbine Installers

There is minimal research published concerning wind turbine installers. The installation work may be completed by local electrical contractors under the strict guidelines of the project manager. The Irish Wind Energy Association (IWEA) (Irish Wind Energy Association 2016) provides many services on their website among which is a directory of wind turbine installers. The contact details are listed for each of these installers. The IWEA is a Non-Governmental-Organisation (NGO) who claims to be committed to the promotion and education of wind energy issues. They are also a lobbying group who work to influence government policy on renewable energy. The IWEA is linked professionally to the European Wind Energy Association. The equivalent association in the UK is the British Wind Energy Association (British Wind Energy Association 2016).

Table 2.1 Global wind turbine manufacturers

Siemens	Market leader in offshore wind turbines. Siemens turbines have been running for more than 20 years at Vinneby, Denmark, the world's first offshore wind farm. Siemens has approximately 348,000 employees in over 200 countries (Siemens 2016)
GE	US manufacturer. GE Energy's 1.5-MW series is the most widely deployed wind turbine, with more than 16,000 installed across the globe (GE Renewable Energy 2016)
Vestas	Focused completely on the wind industry, based in Denmark. They have been in the wind industry since the 1970s, and there are now more than 43,000 Vestas turbines installed in 66 countries. Vestas has offices in 24 countries (Vestas 2016)
Goldwind	Recently emerged as China's leading turbine manufacturer. Goldwind claims to be the worlds' leading manufacturer of permanent magnet direct-drive turbines. They produce 1.5-MW and 2.5-MW wind turbine models (Goldwind 2016)
Enercon	The company was found in 1984 as a dedicated wind turbine manufacturer. Enercon now has more than 20,000 turbine installations around the world. The German company depends heavily on the German market (Enercon 2016)
Gamesa	Spanish manufacturer with a strong presence in India and Latin America. They put a large focus on the servicing industry. They now have manufacturing plants in the USA, Brazil, China, and India (Gamesa 2016)
United Power	This company is part of one of China's biggest power producers, China Guodian (United Power 2016)
Ming Yang	This company is based in China, but unlike the other Chinese wind turbine manufacturers does not benefit from being state-owned. The company was established in 2007. The company is heavily dependent on European technology and expertise (Ming Yang 2016)
Senvion	German turbine manufacturer formerly called Repower. They employ approximately 3500 people in 14 countries (Senvion 2016)
Nordex	Based in Germany, Nordex is one of the pioneers in wind turbine technology. It produced the first megawatt-size turbine in 1995 with the N54—1 MW model (Nordex 2016)

2.5.4 Other Renewable Energy Technologies

While wind projects made the most significant renewable energy contribution in Ireland in 2015, Table 2.2, other renewable energy sources available for use include Bioenergy, Solar, Hydro, and Ocean energy sources. Table 2.2 displays the sources of renewable electricity by technology and their installed capacity and contribution to gross electricity consumption in Ireland in 2015 (SEAI 2016). Poor decision-making manifests itself in the lack of exploring suitable alternatives (Tylock et al. 2012) so perhaps Solar and Hydro could be explored and researched to increase their penetration.

The Biomass and Renewable Waste contribution of 1.0% includes a small contribution of solid Biomass CHP (Combined-Heat-Power). The overall share of electricity from renewable energy increased significantly between 1990 and 2015 with

Table 2.2 Renewable energy sources by technology in Ireland in 2015 (SEAI 2016)

Renewable technology	% of gross electricity (normalised)
Hydro	2.5
Wind	21.1
Biomass & Renewable waste	1.0
Landfill gas	0.6
Biogas	0.1
Solar	0.01
Total	25.3

a value of 5.3% in 1990 and 25.3% (normalised) in 2015. Biogas consists of land-fill gas, sewage sludge gas, and other biogas produced by anaerobic digestion. The Landfill Gas contribution is reported separately, as shown in Table 2.2. SEAI warns that the value of the Biogas contribution (Wastewater Treatment plants and other Bio-gas installations in industry) to the Irish national energy balance shown in Table 2.2 is mostly an estimated figure. This estimation is due to the poor response rates to the SEAI annual surveys. The SEAI state that there are currently few grid-connected photovoltaic (PV) installations in Ireland although interest in the technology is grow-ing. They also state that solar technology is one of the technologies being considered in the context of the new, 2017, support scheme for renewable electricity generation. Ocean energy (Tidal and Wave) is in the Research and Development (R&D) stage and is expected to make some contribution in the future.

2.6 Discussion and Conclusions

This literature review concurred with research by Danilovic et al. (2015) who found that there are an ever-increasing complexity and progression in the CSR-related field. These authors (Danilovic et al. 2015) used the bibliometric method to measure the number of publications and the rate of growth of articles in established databases through which the CSR concept was diffused. The findings of the paper (Danilovic et al. 2015) called for studies within academia to go deeper in attempting to under-stand the CSR concept. In attempting to understand and define CSR, enough evidence was reviewed to advise that CSR and its related concept Sustainable Development can be defined as a management decision-making framework comprised of three dis-tinct yet interconnected components, namely, economic, environmental, and social components (Fig. 2.1). Literature for the Elkington (1997) framework nestles mainly within the discipline of management, but there is some evidence of crossover into other disciplines that warrant business acumens such as engineering and economics. Sustainability/CSR management-based literature can broadly be sorted into five main business management perspectives, namely, marketing, leadership, strategy, stake-holder theory, and social obligation perspectives. There appears to be a renewed focus on the stakeholder theory perspective, as described in Sect. 2.2.1.4.

The review highlighted the ambiguous/subjective nature of CSR/Sustainable Development (Sect. 2.2.2). Karassin and Bar-Haim (2016) claim that one of the reasons for the ambiguity is that there is no single standard measurement system that accurately represents and allows comparison of performance across different industries. In a similar vein, Xu et al. (2016) claim that much of the CSR research has little integrity or universality due to the lack of a systematic analytical framework. The lack of a systematic framework is addressed in this current study. Despite the ambiguity, some research purports to assess the link between CSR activities within an organisation and each of the three component domains (economic, environmental, and human). In the economic domain, literature has been published assessing the link between CSR activities and financial performance of a business with mixed results (Nollet et al. 2016; Wang and Xiao 2016). Much of the literature assessing the environmental benefits of implementing CSR initiatives focus on climate change/global warming phenomenon (Carrington 2016). Responses to the universal call to reduce CO_2 emissions (Bentley 2016) has made the sustainable energy sector more visible, but this author found that literature in the CSR/Environment/Alternative Energy Systems area appears to be confined mainly to commentary and rhetorical pieces rather than significant robust empirical research studies. While there seems to be a substantial impetus to use wind turbines as energy-generating alternatives to fossil-fuel-driven traditional plants both at an Irish (Sect. 1.1.3) and international (Sect. 2.5.1) level the decisions (public and private) to implement such wind turbine technologies to reduce CO_2 emissions have not been fully validated (Kaffine et al. 2013; Trebilcock 2009). Some of the literature claims that problems arise because the intermittent production of wind power requires rapid adjustment of parallel-connected fossil-fuel generators in response to ramping (up and down) in wind generators (Kaffine et al. 2013). Intermittent in this context implies variations in wind speed/power output on an hourly basis (or even on a daily basis). The cycling (ramping) phases can lead to higher than average emission rates from the fossil-fuel back-up generators (Cullen 2013) and therefore negate the projected beneficial effects of the wind turbine generators. Some studies were reviewed that use hourly wind generation data (Kaffine et al. 2013) but no studies were found utilising wind turbine power output data on a significantly shorter time-frame (half-second) to ascertain if the ramping phenomenon is occurring on a more continuous basis. This current study obtained and analysed such half-second data using a power quality measuring tool, the results of which are demonstrated in Chaps. 4 and 6 on a 10-kW three-phase synchronous generator and a 300-kW induction-type generator, respectively.

In the social (human) domain many publications suggest that CSR activities have a positive influence on the attitude of company employees (Dumitrescu and Simionescu 2015; Voegtlin and Greenwood 2016; Lapina et al. 2013). Some authors (Mea and Sims 2019) discuss how religious perspectives can empower business leaders to build an ethos of humanistic management. The authors focus specifically on catholic social doctrine and teaching and explain how the Catholic Church supports the idea that profit and economic efficiency are desirable. The outcome of the three sustainability components (TBL) being beneficial serves one purpose—human flourishing. Human flourishing and sustainability mean a capacity to endure over the long term. Kuhnen and Hahn (2019) suggest that environmental protection and economic prosperity are a means to an end of ensuring human well-being, i.e. the social dimension of sustainability. It appears that the social component of sustainability initiatives/projects is accomplished when the projects are economically viable, *and* the initiatives/projects produce no outcome that can do damage to the environment.

References

Abernathy, J. L., Guo, F., Kubick, T. R., & Masli, A. (2019). Financial statement footnote readability and corporate audit outcomes, *AUDITING: A Journal of Practice & Theory, 38*(2), 1–26. https://doi.org/10.2308/ajpt-52243.

Abro, M. M. Q., Khurshid, M. A., & Aamir, A. (2016). Corporate Social Responsibility (CSR) practices: The case of Saudi Aramco. *Journal of Competitiveness Studies, 24*(1), 79–90.

Agudo-Valiente, J. M., Garces-Ayerbe, C., & Salvador-Figueras, M. (2015, January/February). Corporate social performance and Stakeholder dialogue management. *Corporate Social Responsibility and Environmental Management, 22*(1), 13–31.

Anand, S., & Segal, P. (2008). What do we know about global income inequality? *Journal of Economic Literature, 46*(1), 57–94.

Anderson, T. R., Hawkins, E., & Jones, P. D. (2016, September). CO_2, the greenhouse effect and global warming: From the pioneering work of Arrhenius and Callendar to today's Earth System Models. *Endeavour, 40*(3), 178–187.

Andrade, J. C. S., & Puppim de Oliveira, J. A. (2015, August). The role of the private sector in global climate and energy governance. *Journal of Business Ethics, 130*(2), 375–387.

Atkinson, A. B. (2016, January/February). How to spread the wealth. *Foreign Affairs, 95*(1), 29–33.

Battaglia, M., Passetti, E., Bianchi, L., & Frey, M. (2016). Managing for integration: A longitudinal analysis of management control for sustainability. *Journal of Cleaner Products, Part A, 136*, 213–225.

BBC. http://www.bbc.com. Accessed on 3 November 2016.

Bednar, M. K. (2012). Watchdog or lapdog? A behavioural view of the media as a corporate governance mechanism. *Academy of Management Journal, 55*(1), 131–150.

Benabou, R. (2013). Groupthink: Collective delusions in organisations and markets. *Review of Economic Studies, 80*, 429–462. https://doi.org/10.1093/restud/rds030.

Bentley, Y. (2016, July). UK company strategies in reducing carbon dioxide emissions. *International Journal of Business & Economic Development, 4*(2), 78–86.

Besio, C., & Prozini, A. (2014). Morality, ethics, and values outside and inside organisations: An example of the discourse on climate change. *Journal of Business Ethics, 119*(3), 287–300.

Boiral, O., Henri, J.-F., & Talbot, D. (2012, December). Modeling the impacts of corporate commitment on climate change. *Business Strategy and the Environment, 21*(8), 495–516. https://doi.org/10.1002/bse.723.

Bolis, I., Brunoro, C. M., & Sznelwar, L. I. (2014). Work in corporate sustainability policies: The contribution of ergonomics. *Work, 49*, 417–431. https://doi.org/10.3233/WOR-141962.

Bord Bia. (2015). *Working with nature: An initiative by Bord Bia, Irish Food Board*. Available at http://www.origingreen.ie/. Accessed on 7 March 2015.

Boutsika, T., & Santoso, S. (2013). Quantifying the effect of wind turbine size and technology on wind power variability. *Power and Energy Society General Meeting, IEEE, 2013*, 1–5. https://doi.org/10.1109/PESMG.2013.6672587.

Bowen, H. (1953). *Social responsibility of the businessman*. New York: Harper & Row.

British Wind Energy Association. http://www.bwea.org. Accessed on 2 November 2016.

Brusseau, J., Chiagouris, L., & Fernandez Brusseau, R. (2013, Spring). Corporate social responsibility: To yourself be true. *Journal of Global Business and Technology, 9*(1), 53–63.

Buckley, P. J., & Strange, R. (2015, May). The governance of the global factory: Location and control of world economic activity. *Academy of Management Perspectives, 29*(2), 237–249. https://doi.org/10.5465/amp.2013.0113.

Buhmann, K., Jonsson, J., & Fisker, M. (2019). Do no harm and do more good too: Connecting the SDGs with business and human rights and political CSR theory. *Corporate Governance, 19*(3), 389–403. https://doi.org/10.1108/CG-01-2018-0030.

Buyaert, P. (2012, January). CSR and leadership: Can China lead a new paradigm shift? *Asian Journal of Business Ethics, 1*(1), 73–77, https://doi.org/10.1007/s13520-011-0007-z.

Carrington, D. (2016, November 5 Saturday). Oil firms announce $1bn climate fund to reduce impact of fossil fuels. *The Guardian*, Financial, p. 32.

Carroll, A. B. (1979). A three-dimensional model of corporate performance. *Academic Management Review, 4*(4), 497–505.

Carroll, A. B. (1991, July/August). The Pyramid of corporate social responsibility: Toward the moral management of organisational Stakeholders. *Business Horizons, 34*(4), 39–48.

Carroll, A. B. (1999, September). Corporate social responsibility: Evolution of a definitional construct. *Business & Society, 38*(3), 268–295.

Catalin, C., & Nicoleta, R. (2011). International Biomass trade and sustainable development: An overview. *Annals of the University of Oredea, Economic Science Series, 20*(2), 47–54.

Chapman, A., McLellan, B., & Tezuka, T. (2016, September). Strengthening the energy policy making process and sustainability outcomes in the OECD through policy design. *Administrative Sciences, 6*(3), 1–16. https://doi.org/10.3390/admsci6030009.

Choudhury, A. K., Shankar, R., & Tiwari, M. K. (2006, December). Consensus-based intelligent group decision-making model for the selection of advance technology. *Decision Support Systems, 42*(3), 1776–1799. https://doi.org/10.1016/j.dss.2005.05.001.

Collins Dictionary. http://www.collinsdictionary.com. Accessed on 7 November 2016.

Commission for Energy Regulation. http://www.cer.ie. Accessed on 7 November 2016.

Constantine, T. (2015, January). Social responsibility, quality of work life and motivation to contribute in the Nigerian society. *Journal of Business Ethics, 126*(2), 219–233. https://doi.org/10.1007/s10551-013-1940-7.

Coors, A. C., & Winegarden, W. (2005, Spring). Corporate social responsibility—Or good advertising. *Regulation, 28*(1), 10–11.

COP21. (2015). *Conference of Parties, 2015 United Nations Climate Change Conference*. Paris, 30 November–12 December 2015.

Costa, P. L., Passos, A. M., & Bakker, A. B. (2014). Team work engagement: A model of emergence. *Journal of Occupational and Organisational Psychology, 87*, 414–436.

Coy, P., Parkin, B., & Martin, A. (2014, March 24). In trade talks, its countries vs. companies. *Bloomberg Businessweek*, 4372, pp. 35–37.

Cullen, J. (2013). Measuring the environmental benefits of wind-generated electricity. *American Economic Journal: Economic Policy, 5*(4), 107–133. https://doi.org/10.1257/pol.5.4.107.

Dahan, N., Doh, J., & Raelin, J. (2015, October). Pivoting the role of government in the business and society interface: A Stakeholder perspective. *Journal of Business Ethics, 131*(3), 665–680. https://doi.org/10.1007/s10551-014-2297-2.

Dahlsrud, A. (2008, January/February). How corporate social responsibility is defined: An analysis of 37 definitions. *Corporate Social Responsibility and Environmental Management, 15*(1), 1–13. https://doi.org/10.1002/csr.132.

Danilovic, M., Hensbergen, M., Hoveskog, M., & Zadayannaya, L. (2015, May). Exploring diffusion and dynamics of corporate social responsibility. *Corporate Social Responsibility & Environmental Management, 22*(3), 129–141. https://doi.org/10.1002/csr.1326.

Davis, K. (1960, Spring). Can business afford to ignore its social responsibilities? *California Management Review, 2*(3), 70–76.

De Rezende, L. B., Blackwell, P., & Pessanha Goncalves, M. D. (2018, February/March). Research focuses, trends and major findings on project complexity: A bibliometric network analysis of 50 years of project complexity research. *Project Management Journal, 49*(1), 42–56.

de Vries, G., Terwel, B. W., Ellemers, N., & Daamen, D. D. L. (2015, May). Sustainability or profitability? How communicated motives for environmental policy affect public perceptions of corporate greenwashing. *Corporate Social Responsibility & Environmental Management, 22*(3), 142–154. https://doi.org/10.1002/csr.1327.

Demir, G., Cagle, M. N., & Dalkilic, A. F. (2016). Corporate social responsibility and regulatory initiatives in Turkey: Good implementation examples. *Accounting & Management Information Systems, 15*(2), 372–400.

Devin, B. (2016, March). Half-truths and dirty secrets: Omissions in CSR Communications. *Public Relations Review, 42*(1), 226–228.

Diaz, S., Carta, J. A., & Matias, J. M. (2018, January). Performance assessment of five MCP models proposed for the estimation of long-term wind turbine power outputs at a target site using three machine learning techniques. *Applied Energy, 209*, 455–477. https://doi.org/10.1016/j.apenergy.2017.11.007.

Dragomir, G., Serban, A., Nastase, G., & Brezeanu, A. I. (2016, October). Wind energy in Romania: A review from 2009 to 2016. *Renewable and Sustainable Energy Reviews, 64*, 129–143.

Dumitrescu, D., & Simionescu, L. (2015). Empirical research regarding the influence of corporate social responsibility activities on companies' employees and financial performance. *Economic Computation & Economic Cybermetrics Studies and Research, 49*(3), 52–66.

Eirgrid. (2015). *Connected TSO non-wind generators 18th September 2015*. Available from http://www.eirgridgroup.com. Accessed on 8 December 2015.

Element Power. http://www.elpower.com. Accessed on 2 November 2016.

Elkington, J. (1997). *Cannibal with forks: The triple bottom line of 21st century business*. Gabriola Island, BC: Capstone.

Enercon. http://www.enercon.de/en/home/. Accessed on 8 November 2016.

Foellmi, R., and Oechslin, M. (2010, May). Market imperfections, wealth inequality, and the distribution of trade gains. *Journal of International Economics, 81*(1), 15–25.

Fremeaux, S., & Michelson, G. (2017). The common good of the firm and humanistic management: Conscious capitalism and economy of communion. *Journal of Business Ethics* (145), 701–709. https://doi.org/10.1007/s10551-016-3118-6.

Frias-Aceituno, J. V., Rodriguez-Ariza, L., & Garcia-Sanchez, I. M. (2013). The role of the board in the dissemination of integrated corporate social reporting. *Corporate Social Reporting and Environmental Management, 20*, 219–233. https://doi.org/10.1002/csr.1294.

Friedman, M. (1970, September 13). The social responsibility of business is to increase its profits. *New York Times*, pp. 122–126.

Frostenson, M. (2016, September). Humility in business: A contextual approach. *Journal of Business Ethics, 138*(1), 91–102. https://doi.org/10.1007/s10551-015-2601-9.

Gaelectric. http://www.gaelectric.ie. Accessed on 2 November 2016.

Galupa, A., Hartulari, C., & Spataru, S. (2014). The environment pollution in terms of system theory and multicriterial decisions. *Economic Computation & Economic Cybernetics Studies & Research, 48*(4), 175–189.

Gamesa. http://www.gamesacorp.com/en/. Accessed on 8 November 2016.

Garen, J. (2010). On fairness and needs in a free enterprise economy. *Journal of Applied Economics and Policy, 29*(1), 61–78.

GE Renewable Energy. https://www.gerenewableenergy.com/wind-energy/turbines.html. Accessed on 8 November 2016.

Gerard, J. A., & Weber, C. M. (2015, June). Compliance and corporate governance: Theoretical analysis of the effectiveness of compliance based on locus of functional responsibility. *International Journal of Global Business, 8*(1), 15–26.

Ghoseiri, K., & Lessan, J. (2014). Waste disposal site selection using an analytic hierarchal pairwise comparison and ELECTRE approaches under fuzzy environment. *Journal of Intelligent & Fuzzy Systems, 26*(2), 693–704. https://doi.org/10.3233/IFS-120760.

Glavas, A., & Mish, J. (2015, March). Resources and capabilities of triple bottom line firms: Going over old or breaking new ground? *Journal of Business Ethics, 127*(3), 623–642. https://doi.org/10.1007/s10551-014-2067-1.

Global Wind Energy Council. http://www.gwec.net. Accessed on 2 November 2016.

Goldwind. http://www.goldwindglobal.com/web/index.do. Accessed on 8 November 2016.

Gonzalez, J. S., & Lacal-Arantegui, R. (2016, April). A review of regulatory framework for wind energy in European Union Countries: Current state and expected developments. *Renewable and Sustainable Energy Reviews, 56*, 588–602.

Gonzalez-Rodriguez, M. D. R., Del Carmen Diaz-Fernandez, M., Spers, V. R. E., & Da Silva Leite, M. (2016, January/February). Relation between background variables, values and corporate social responsibility. *Revista de Administracao de Empresas, 56*(1), 8–19. https://doi.org/10.1590/s0034-759020160102.

Gutierrez, R., & Vernis, A. (2016, June). Innovations to serve low-income citizens: When corporations leave their comfort zones. *Long Range Planning, 49*(3), 283–297.

Hack, L., Kenyon, A. J., & Wood, E. H. (2014). A critical Corporate Social Responsibility (CSR) timeline: How should it be understood now? *International Journal of Management Cases, 16*(4), 46–55.

Hansen, E. G., Zvezdov, D., Harms, D., & Lenssen, G. (2014). Advancing corporate sustainability, CSR, and business ethics. *Business & Professional Ethics Journal, 33*(4), 287–296.

He, Z.-X., Xu, S.-C., Shen, W.-X., Long, R.-Y., & Chen, H. (2016, October). Factors that influence corporate environmental behaviour: Empirical analysis based on panel data in China. *Journal of Cleaner Production, 133*, 531–543.

Heaslip, E., Costello, G. J., & Lohan, J. (2016). Assessing good-practice frameworks for the development of sustainable energy communities in Europe: Lessons from Denmark and Ireland. *Journal of Sustainable Development of Energy, Water and Environmental Systems, 4*(3), 307–319. https://doi.org/10.13044/j.sdewes.2016.04.0024.

Herremans, I. M., Nazari, J. A., & Mahmoudian, F. (2016). Stakeholder relationships, engagement, and sustainability. *Journal of Business Ethics, 138*(3), 417–435. https://doi.org/10.1007/s10551-015-2634-0.

Higgins, C., Stubbs, W., & Milne, M. (2018). Is sustainability reporting becoming institutionalised? The role of an issues-based field. *Journal of Business Ethics, 147*(2), 309–326. https://doi.org/10.1007/s10551-015-2931-7.

Holme, L., & Watts, P. (2000). *Corporate social responsibility, making good business sense.* Geneva: World Business Council for Business Development.

Howard, J. A. (1950, January). New British Law on Monopoly. *Journal of Marketing, 14*(4), 590–594.

Hrasky, S. (2012). Carbon footprints and legitimation strategies: Symbolism or action? *Accounting, Auditing & Accountability Journal, 25*(1), 174–198.

Iglinski, B., Iglinska, A., Kozinski, G., Skrzatek, M., & Buczkowski, R. (2016, October). Wind energy in Poland—History, current state, surveys, renewable energy sources act, SWOT analysis. *Renewable and Sustainable Energy Reviews, 64*, 19–33.

IPCC. (2013). *Intergovernmental panel on climate change, summary for policymakers.* Available from https://www.ipcc.ch/pdf/assessment-report/ar5/wg1/WG1AR5_SPM_FINAL.pdf. Accessed on 16 January 2016.

Irish Wind Energy Association. http://www.iwea.com. Accessed on 2 November 2016.

Janis, I. (1972). *Victims of groupthink: Psychological studies of policy decisions and fiascos.* Boston, MA: Houghton Mifflin Company.

Jia, M., Tong, L., Viswanath, P. V., & Zhang, Z. (2016, October). Word power: The impact of negative media coverage on disciplining corporate pollution. *Journal of Business Ethics, 138*(3), 437–458. https://doi.org/10.1007/s10551-015-2596-2.

Jonczyk, C. D., Lee, Y. G., Galunic, C. D., & BensaouInsead, B. M. (2016, June). Relational changes during role transitions: The interplay of efficiency and cohesion. *Academy of Management Journal, 59*(3), 956–982. https://doi.org/10.5465/amj.2013.0972.

Jones, G., & Bouamane, L. (2011, May 4). *Historical trajectories and corporate competences in wind energy* (Working Papers—Harvard Business School Division of Research, Paper No 11–112). https://doi.org/10.2139/ssm.1831471.

Jung, H.-J., & Kim, D.-O. (2016). Good neighbours but bad employers: Two faces of corporate social responsibility programmes. *Journal of Business Ethics, 138*(2), 295–310. https://doi.org/10.1007/s10551-015-2587-3.

Kaffine, D. T., McBee, B. J., & Lieskovsky, J. (2013). Emissions savings from wind power generation in Texas. *The Energy Journal, 34*(1), 155–175. https://doi.org/10.5547/01956574.34.1.7.

Kang, Y. J. (2019, August). Are audit committees more challenging given a specific investor base? Does the answer change in the presence of prospectice critical audit matter disclosures? *Accounting, Organisations, and Society, 77*, 1–14. https://doi.org/10.1016/j.aos.2019.04.001.

Karaibrahimoglu, Y. Z., & Cangarli, B. G. (2016). Do auditing and reporting standards affect firms' ethical behaviours? The moderating role of national culture. *Journal of Business Ethics, 139*(1), 55–75. https://doi.org/10.1007/s10551-015-2571-y.

Karassin, O., & Bar-Haim, A. (2016, December). Multilevel corporate environmental responsibility. *Journal of Environmental Management, 183*, Part 1, pp. 110–120.

Katzenstein, W., & Apt, J. (2009). Air emissions due to wind and solar power. *Environmental Science and Technology, 43*(2), 253–258. https://doi.org/10.1021/es801437t.

Kealy, T. (2014, April). Financial appraisal of a small scale wind Turbine with a case study in Ireland. *Journal of Energy and Power Engineering, 8*(4), 620–627. https://doi.org/10.17265/1934-8975/2014.04.004.

Kealy, T. (2016, October). Impediments in the CSR space: A mixed-method approach. *International Journal of Advanced Research, 4*(10), 1995–2026. https://doi.org/10.21474/ijar01/2015.

Kealy, T. (2019). Triple bottom line sustainability reporting: How to make it more tangible. *American Journal of Management, 19*(5), 107–140.

Kealy, T., Barrett, M., & Kearney, D. (2015, April). How profitable are wind turbine projects? An empirical analysis of a 3.5 MW wind farm in Ireland. *International Journal on Recent Technologies in Mechanical and Electrical Engineering, 2*(4), 58–63.

Kearns, A. J. (2017). Rebuilding trust: Ireland's CSR plan in the light of Caritas in Veritate. *Journal of Business Ethics, 146*(4), 845–857. https://doi.org/10.1007/s10551-016-3238-z.

Kelman, S., Sanders, R., & Pandit, G. (2017). "Tell it like it is": Decision making, groupthink, and decisiveness among US federal subcabinet executives. *Governance: An International Journal of Policy, Administration, and Institutions, 30*(2), 245–261. https://doi.org/10.1111/gove.12200.

Kinley, D. (Ed.). (2017). *Human rights and corporations*. New York: Routledge. ISBN 9780754627425 (hbk).

Kolmar, M., & Beschorner, T. (2016). Locating responsibility: An extended transaction cost approach. *ZeitschriftfuerWirtschafts und Unternehmensethik, 17*(1), 118–147.

Kose, F., Aksoy, M. H., & Ozgoren, M. (2014). An assessment of wind energy potential to meet electricity demand and economic feasibility in Konya, Turkey. *International Journal of Green Energy, 11*(6), 559–576.

Krisnawati, A., Yudoko, G., & RosBangun, Y. (2014). Development path of corporate social responsibility theories. *World Applied Sciences Journal*, 110–120, IDOSI Publications.

Kuhnen, M., & Hahn, R. (2019). From SLCA to positive sustainability performance measurement: A two-tier Delphi study. *Journal of Industrial Ecology, 23*(3), 615–634. https://doi.org/10.1111/jiec.12762.

Kurucz, E. C., Colbert, B. A., Ludeke-Freund, F., Upward, A., & Willard, B. (2017, January). Relational leadership for strategic sustainability: Practices and capabilities to advance the design and assessment of sustainable business models. *Journal of Cleaner Production, 140*, Part 1, 189–204.

Lakhanpal, S. (2019). Contesting renewable energy in the global south: A case-study of local opposition to a wind power project in the Western Ghats of India. *Environmental Development, 30*(6), 51–60. https://doi.org/10.1016/j.envdev.2019.02.002.

Lam, H. K. S., Yeung, A. C. L., Cheng, T. C. E., & Humphreys, P. K. (2016, October). Corporate environmental initiatives in the Chinese context: Performance implications and contextual factors. *International Journal of Production Economics, 180*, 48–56.

Langer, K., Decker, T., Roosen, J., & Menrad, K. (2016, October). A qualitative analysis to understand the acceptance of wind energy in Bavaria. *Renewable and Sustainable Energy Reviews, 64*, 248–269.

Langer, K., Decker, T., Roosen, J., & Menrad, K. (2018, February). Factors influencing citizens' acceptance and non-acceptance of wind energy in Germany. *Journal of Cleaner Production*, 175, 133–144. https://doi.org/10.1016/j.jclepro.2017.11.221.

Lapina, I., Maurane, G., & Starineca, O. (2013). Human resource management models: Aspects of knowledge management and corporate social responsibility. *Procedia—Social and Behavioural Sciences*, January 2014, *110*, 577–586.

Lavoie, M. (2015). The Eurozone crisis: A balance-of-payments problem or a crisis due to a flawed monetary design? *International Journal of Political Economy, 44*(2), 157–160. https://doi.org/10.1080/08911916.2015.1060831.

Lock, I., & Seele, P. (2016, October). Deliberate lobbying? Toward a noncontradiction of corporate political activities and corporate social responsibility? *Journal of Management Inquiry, 25*(4), 415–430.

Lyon, T., Lu, Y., Shi, X., & Yin, Q. (2013, October). How do Investors respond to Green Company awards in China? *Ecological Economics, 94*, 1–8. https://doi.org/10.1016/j.ecolecon.2013.06.020.

Maak, T., Pless, N. M., & Voegtlin, C. (2016, May). Business statesman or shareholder advocate? CEO responsible leadership styles and the micro-foundations of political CSR. *Journal of Management Studies, 53*(3), 463–493.

Maas, K., Schaltegger, S., & Crutzen, N. (2016, November). Integrating corporate sustainability assessment, management accounting, control, and reporting. *Journal of Cleaner Production*, Part A, *136*, 237–248.

MacCormac, S., & Haney, H. (2012, Spring). One viable solution to advancing environmental sustainability. *Journal of Applied Corporate Finance, 24*(2), 49–56.

Madorran, C., & Garcia, T. (2016, January/February). Corporate social responsibility and financial performance: The Spanish case. RAE—*Revista de Administracao de Empresas, 56*(1), 20–28. https://doi.org/10.1590/s0034-759020160103.

Mahoney, L. S., Thorne, L., Cecil, L., & LaGore, W. (2013, June). A research note on standalone corporate social responsibility reports: Signaling or greenwashing? *Critical Perspectives on Accounting, 24*(4–5), 350–359.

Mainstream Renewable Power. http://www.mainstreamrp.com. Accessed on 2 November 2016.

Maroun, W. (2019). Exploring the rationale for integrated report assurance. *Accounting, Auditing & Accountability Journal, 32*(6), 1826–1854. https://doi.org/10.1108/AAAJ-04-2018-3463.

Matten, D., & Crane, A. (2005, January). Corporate citizenship: Toward an extended theoretical conceptualisation. *The Academy of Management Review, 30*(1), 166–179.

Matuleviciene, M., & Stravinskiene, J. (2015). The importance of Stakeholders for corporate reputation. *Engineering Economics, 26*(1), 75–83. https://doi.org/10.5755/j01.ee.26.1.6921.

Mayer, S. E., Lopoo, L. M., & Groves, L. H. (2016, October). Government spending and the distribution of economic growth. *Southern Economic Journal, 83*(2), 299–415. https://doi.org/10.1002/soej.12148.

Mea, W. J., & Sims, R. R. (2019). Human dignity-centered business ethics: A conceptual framework for business leaders. *Journal of Business Ethics, 160*(1), 53–69. https://doi.org/10.1007/s10551-018-3929-8.

Mehdi, B. (2015, October). Control in the multinational enterprise: The polycentric case of global professional service firms. *Journal of World Business, 50*(4), 696–703.

Melia, P. (2015, November 24 Tuesday). Critics of wind energy 'stoke fake technology fears' despite success; Debate is needed to allay public concerns. *Irish Independent*, News, p. 19.

Mercier, G., & Deslandes, G. (2017). There are no codes, only interpretations. Practical Wisdom and Hermeneutics in Monastic Organisations. *Journal of Business Ethics* (145), 781–794. https://doi.org/10.1007/s10551-016-3055-4.

Midttun, A., Gjolberg, M., Kourula, A., Sweet, S., & Vallentin, S. (2015, July). Public policies for corporate social responsibility in four Nordic countries: Harmony of goals and conflict of means. *Business & Society, 54*(4), 464–500.

Miles, S. (2017, May). Stakeholder theory classification: A theoretical and empirical evaluation of definitions. *Journal of Business Ethics, 142*(3), 71–98. https://doi.org/10.1007/s10551-015-2741-y.

Milne, M. J., & Gray, R. (2013). W(h)ither ecology? The triple bottom line, the global reporting initiative, and corporate sustainability reporting. *The Journal of Business Ethics, 118,* 13–29.

Ming Yang. http://www.mywind.com.cn/English/index.aspx. Accessed on 8 November 2016.

Mishra, S., & Modi, S. B. (2016, January). Corporate social responsibility and shareholder wealth: The role of marketing capability. *Journal of Marketing, 80*(1), 26–46. https://doi.org/10.1509/jm.15.0013.

Nikas, A., Doukas, H., & Martinez Lopez, L. (2018, March). A Group decision making tool for assessing climate policy risks against multiple criteria. *Heliyon, 4*(3). https://doi.org/10.1016/j.heliyon.2018.e00588.

Nollet, J., Filis, G., & Mitrokostas, E. (2016, January). Corporate social responsibility and financial performance: A non-linear and disaggregated approach. *Economic Modelling, 52*, Part B, 400–407.

Nordex. http://www.nordex-online.com/en. Accessed on 8 November 2016.

Okyhusen, G. A., Lepak, D., Ashcraft, L., Labianca, G., Smith, V., & Steensma, H. K. (2015). Theories of work and working today. *Academy of Management Review*, 2015 Supplement, *4015*, 6–17. https://doi.org/10.5465/amr2013.0169.test.

Osagie, E., Wesselink, R., Blok, V., Lans, T., & Mulder, M. (2016, May). Individual competencies for corporate social responsibility: A literature and practice perspective. *Journal of Business Ethics, 135*(2), 233–252. https://doi.org/10.1007/s10551-014-2469-0.

Paredes-Gazquez, J. D., Rodriguez-Fernandez, J. M., & Cuesta-Gonzalez, M. (2016). Measuring corporate social responsibility using composite indices: Mission impossible? The Case of the Electricity Utility Industry. *Spanish Accounting Review, 19*(1). https://doi.org/10.1016/j.rcsar.2015.10.001.

Perego, P., & Kolk, A. (2012, October). Multinationals' accountability on sustainability: The evolution of third-party assurance of sustainability reports. *Journal of Business Ethics, 110*(2), 173–190. https://doi.org/10.1007/s10551-012-1420-5.

Pinkse, J., & Kolk, A. (2010, May). Challenges and trade-offs in corporate innovation for climate change. *Business Strategy and the Environment, 19*(4), 261–272.

Pope Francis. (2015). Laudato Si'. *Encyclical letter of the Holy Father Francis on care for our common home*. Veritas.

Prado-Lorenzo, J.-M., & Garcia-Sanchez, I.-M. (2010, December). The role of the board of directors in disseminating relevant information on greenhouse gases. *Journal of Business Ethics, 97*(3), 391–424.

Ptak, T. (2019). Towards an ethnography of small hydropower in China: Rural electrification, socioeconomic development and furtive hydroscapes. *Energy Research and Social Science, 48*(2), 116–130. https://doi.org/10.1016/j.erss.2018.09.010.

Puga, J. (2010). The importance of combined cycle generating plants in integrating large levels of wind power generation. *The Electricity Journal, 23*(7), 33–44.

Ramlugun, V. G., & Raboute, W. G. (2015). Do CSR practices of banks in mauritius lead to satisfaction and loyalty? *Studies in Business & Economics, 10*(2), 128–144. https://doi.org/10.1515/sbe-2015-0025.

Rasche, A., & Esser, D. E. (2006). From stakeholder management to stakeholder accountability—Applying Habermasian discourse ethics to accountability research. *Journal of Business Ethics, 65*(3), 251–267.

Reich, R. B. (2009, July–August). Government in your Business. *Harvard Business Review, 87*(7/8), 94–99.

Riccobono, F., Bruccoleri, M., & Grobler, A. (2016, April). Groupthink and project performance: The influence of personal traits and interpersonal ties. *Production and Operations Management, 25*(4), 609–629. https://doi.org/10.1111/poms.12431.

Rivera, J. M., Munoz, M. J., & Moneva, J. M. (2017, November/December). Revisiting the relationship between corporate stakeholder commitment and social and financial performance. *Sustainable Development, 25*(6), 482–494. https://doi.org/10.1002/sd.1664.

Saeidi Pour, B., Nazari, K., & Emami, M. (2014, April). Corporate social responsibility: A literature review. *African Journal of Business Management, 8*(7), 228–234. https://doi.org/10.5897/ajbm12.106.

Sandelands, L. E., & Hoffman, A. J. (2008). *Sustainability, Faith, and the market.* Working Paper (Faculty), University of Michigan Business School, p. 1.

Santos, R. F., Antunes, P., Ring, I., & Clemente, P. (2015, March/April). Engaging local private and public actors in biodiversity conservation: The role of agri-environmental schemes and ecological fiscal transfers. *Environmental Policy & Governance, 25*(2), 83–96. https://doi.org/10.1002/eet.1661.

Schaltegger, S., & Burritt, R. (2018, January). Business cases and corporate engagement with sustainability: Differentiating ethical motivations. *Journal of Business Ethics, 147*(2), 241–259. https://doi.org/10.1007/s10551-015-2938-0.

Schrempf-Stirling, J. (2018, June). State power: Rethinking the role of the state in political corporate social responsibility. *Journal of Business Ethics, 150*(1), 1–14. https://doi.org/10.1007/s10551-016-3198-3.

Schyns, B., & Schilling, J. (2013, February). How bad are the effects of bad leaders? A meta-analysis of destructive leadership and its outcomes. *The Leadership Quarterly, 24*(1), 138–158.

SEAI. (2016, August). *Renewable electricity in Ireland 2015, 2016 report.* Prepared by the Energy Policy Statistical Support Unit. http://www.seai.ie. Accessed on 2 November 2016.

Searcy, C., Dixon, S. M., & Neumann, W. P. (2016, January). The use of work environment performance indicators in Corporate Social Responsibility reporting. *Journal of Cleaner Production, 112*, Part 4, 2907–2921.

Senvion. https://www.senvion.com/global/en/. Accessed on 8 November 2016.

Shakeel, S. R., Takala, J., & Shakeel, W. (2016, October). Renewable energy sources in power generation in Pakistan. *Renewable and Sustainable Energy Reviews, 64*, 421–434.

Siemens. http://www.siemens.com/global/en/home/markets/wind/turbines.html. Accessed on 8 November 2016.

Sims, R. R., & Sauser, W. I. (2013). Towards a better understanding of the relationships among received wisdom, groupthink, and organisational ethical culture. *Journal of Management Policy and Practice, 14*(4), 75–90.

Single Energy Market Operator. http://www.sem-o.com/pages/default.aspx. Accessed on 7 November 2016.

Staniskiene, E., & Stankeviciute, Z. (2018, July). Social sustainability measurement framework: The case of employee perspective in a CSR-committed organisation. *Journal of Cleaner Production, 188*, 708–719. https://doi.org/10.1016/j.jclepro.2018.03.269.

Steinmeier, M. (2016). Fraud in sustainability departments? An exploratory study. *Journal of Business Ethics, 138*(3), 477–492.

Sustainable Energy Authority Ireland. (2017, December). *Energy in Ireland 1990–2016 (2017 Report).* Available at https://www.seai.ie/resources/publications/Energy-in-Ireland-1990-2016-Full-report.pdf. Accessed 12 December 2017.

Sustainable Energy Authority of Ireland. (2015). *Energy in Ireland 1990–2014, 2015 Report.* Available from http://www.seai.ie. Accessed on 8 December 2015.

Sustainable Energy Authority of Ireland, SEAI. (2012, June). The case for sustainable energy, A review and analysis of the economic and enterprise benefits. Version 1.0.

Talbot, D., & Boiral, O. (2018). GHG reporting and impression management: An assessment of sustainability reports from the energy sector. *Journal of Business Ethics, 147*(2), 367–383. https://doi.org/10.1007/s10551-015-2979-4.

The Guardian. (2016, September 17 Saturday). US Business warns against Apple tax fine. *The Guardian*, Financial, p. 36.

Thijssens, T., Bollen, L., & Hassink, H. (2015, December). Secondary Stakeholder influence on CSR disclosure: An application of Stakeholder Salience theory. *Journal of Business Ethics, 132*, 873–891. https://doi.org/10.1007/s10551-015-2623-3.

Trebilcock, M. (2009). *Speaking truth to wind power* (Vol. 22, p. 209). CD Howe Institute.

Tsurumi, T., & Managi, S. (2014, October). The effect of trade openness on deforestation: Empirical analysis for 142 countries. *Environmental Economics & Policy Studies, 16*(4), 305–324. https://doi.org/10.1007/s10018-012-0051-5.

Tylock, S. M., Seager, T. P., Snell, J., Bennett, E. R., & Sweet, D. (2012, August). Energy management under policy and technology uncertainty. *Energy Policy, 47*, 156–163. https://doi.org/10.1016/j.enpol.2012.04.040.

Ullah, A. (2016, April). The impact of free market economy or neoliberalism on labour: Evidence from the garment export sector in Bangladesh. *Middle East Journal of Business, 11*(2), 38–44.

United Nations. (2017). Available at http://www.un.org. Accessed on 4 November 2017.

United Power. http://www.gdupc.com.cn/defaultEn.aspx. Accessed on 8 November 2016.

Van Der Ploeg, L., & Vanclay, F. (2013, September). Credible claim or corporate Spin? A checklist to evaluate corporate sustainability reports. *Journal of Environmental Assessment Policy and Management, 15*(3), 21 pages.

Venter, E., Turyakira, P., & Smith, E. E. (2014, December). The influence of potential outcomes of Corporate Social Responsibility engagement factors on SME's competitiveness. *South African Journal of Business Management, 45*(4), 33–43.

Vestas. https://www.vestas.com/. Accessed on 8 November 2016.

Voegtlin, C., & Greenwood, M. (2016, September). Corporate social responsibility and human resource management: A systematic review and conceptual analysis. *Human Resource Management Review, 26*(3), 181–197.

Wang, S. (2015). Literature review of corporate social responsibility. *Chinese Strategic Decision-making on CSR*, ISBN 978-3-662-44996-7. https://doi.org/10.1007/978-3-662-44997-4-2.

Wang, S., & Xiao, X. (2016). *CSR, strategy, and financial performance, 2016 International Conference on Industrial Economics System and Industrial Security Engineering (IEIS)*. IEEE Conference Publications, 1–6. https://doi.org/10.1109/ieis.2016.7551872.

Warren, C. R., Lumsden, C., O'Dowd, S., & Birnie, R. V. (2005, November). Green on Green: Public perceptions of wind power in Scotland and Ireland. *Journal of Environmental Planning & Management, 48*(6), 853–875. https://doi.org/10.1080/09640560500294376.

Waters, J. (2015, April 5). Nuance or subtlety unwelcome at an RTE that's still in Groupthink's Grip. *Sunday Independent*, News, p. 23.

Wearden, G. (2015, November 4 Wednesday). Volkswagen crisis: German government urges car maker to clear scandal up—Live updates. *The Guardian*.

Webb, A., & Satariano, A. (2016, August 30). Apple Ireland ruling could be the end of easy European Tax deals. *Bloomberg*. Updated 31 August 2016. http://www.bloomberg.com/news/articles/2016-08-30/apple-ireland-ruling-heralds-twilight-of-easy-european-tax-deals. Accessed on 19 September 2016.

Weber, C. M., & Gerard, J. A. (2014). Reconfiguring compliance for corporate social responsibility. *Journal of Economic Development, Management, IT, Finance & Marketing, 6*(2), 28–36.

Wen, J., Hao, Y., Feng, G.-F., & Chang, C.-P. (2016, June). Does government ideology influence environmental performance? Evidence based on a new dataset. *Economic Systems, 40*(2), 232–246. https://doi.org/10.1016/j.ecosys.2016.04.001.

Whaley, L., & Weatherhead, E. (2016, October). Managing water through change and uncertainty: Comparing lessons from the adaptive co-management literature to recent policy developments in England. *Journal of Environmental Planning & Management, 59*(10), 1775–1794.

Wind Power Monthly. 2016. http://www.windpowermonthly.com. Accessed on 5 September 2016.

Wolsink, M., & Breukers, S. (2010, July). Contrasting the core beliefs regarding the effective implementation of wind power. An international study of stakeholder perspectives. *Journal of Environmental Planning and Management, 53*(5), 535–558. https://doi.org/10.1080/09640561003633581.

Xu, J., Li, L., & Zheng, B. (2016, June). Wind energy generation technological paradigm diffusion. *Renewable and Sustainable Energy Reviews, 59*, 436–449.

Xu, A. J., Loi, R., & Lam, L. W. (2015, October). The bad boss takes it all: How abusive supervision and leader-member exchange interact to influence employee silence. *The Leadership Quarterly, 26*(5), 763–774.

Chapter 3
An Irish Perspective on Sustainable Development

Abstract This chapter investigates the current status of sustainable development from an Irish perspective using qualitative analysis on data obtained from semi-structured interviews. The thirteen interviewees were chosen to represent a broad spectrum of influential organisations in Irish society and included private businesses, charity organisations, faith-based organisations, agricultural associations, and government representatives. Analysis of the findings confirmed that the three well-established vital elements that must be considered if organisations are to be developed in a sustainable way, namely, Profit, People, and Planet are still valid. However, this research also found that there were other elements necessary for developing sustainable organisations. To this end, a unique, sustainable organisational development theory emerged from the research findings of this study. This theoretical model identifies four specific vital components that are essential inputs to developing sustainable organisations. These four input components were deemed by the research subjects as critical elements in the ethos and policies of their organisations to maintain sustainability. They included a for-profit ethos, a not-for-profit ethos, an ethical/moral/spiritual ethos and government policy. The study found that organisational leaders guided by a robust moral/ethical compass who incorporate the input components of this model into their organisational vision and strategy maintain a strong focus on sustainable development.

3.1 Introduction

The area of sustainability appears to have gathered pace, with a specific focus on how organisations and specifical businesses can be managed and marketed sustainably. Although some academic research has explored the characteristics and dynamics of sustainable business development, minimal analysis has been carried out in an Irish context. This qualitative study aims to fill the gap and contribute to knowledge in this area. The literature review uncovered a plethora of non-European-based articles, for example, Eweje (2011). The main contribution of this paper is to assess the efforts and perceptions of sustainability in the context of businesses operating currently in Ireland. The initial literature review is followed by a justification of the preferred

© Springer Nature Switzerland AG 2020

T. Kealy, *Evaluating Sustainable Development and Corporate Social Responsibility Projects*, https://doi.org/10.1007/978-3-030-38673-3_3

methodology. The next section presents the findings after which the results are analysed. A discussion section is followed by conclusions, and finally, some suggestions are made to guide future research in this vital area.

3.2   Literature Review

There has been much written on the topic of sustainability in business over the years. It was found that many papers are rhetorical, and opinion based but some research-based material was uncovered. Much of the literature on sustainable business (organisational) development appears to focus on three key areas, economic, environmental, and human aspects within an organisation. Of significance, it was also noted that prominent among the literature surrounding sustainability issues were government/organisational reviews and publications. The extent to which these global organisational reports/studies have impacted on sustainability in business is somewhat questionable.

A definition of sustainability that is often referenced in literature is one from the Bruntland Commission Report of 1987, which states

> Development that meets the needs of the present without compromising the ability of future generations to meet their own needs.

This definition emerged from the Bruntland Commission, which was setup by the United Nations in 1983. Their mission was to direct sustainable development on a global level. This process of globalisation would open up new unprecedented opportunities of large-scale redistribution of wealth, and in such actions, humanity itself would become increasingly interconnected. Whether or not their mission has been realised is far from clear.

3.2.1   Global Context

Among the most pertinent publications on sustainability launched at the start of this new millennium has been the United Nations Global Compact (2014) initiative which included over 12,000 signatories and based in 145 countries. Those corporations that sign up to this pact must adhere to principles in the four main areas of Human Rights, Labour, Environment, and Anti-Corruption. Companies who wish their operations and strategic sustainability efforts to be voluntarily aligned to this international movement can do so with this United Nations Global Compact. Following on from this, a Global Corporate Sustainability Report (2013) was issued. This report assessed the state of corporate sustainability, looking at policies and practices put in place by companies across the range of issues that define a comprehensive approach to responsible business today. Four key findings from this study emerged, briefly described as follows:

(i)   Companies are moving away from good intentions to significant actions,
(ii) Large companies still lead the way,
(iii) Supply chains are a roadblock to improved performance,
(iv) Companies see the big sustainability picture.

The findings point to a clear gap between the 'say' and 'do' steps of the Global Compact Management Model. In particular, Small and Medium-sized Enterprises (SME) find the move from commitment to action more challenging than the large companies. Large companies increasingly view sustainability as a strategic issue and are using the Global Compact principles to help prioritise their responsibility efforts. They are making financial and human resource investments in sustainability. SME cite a lack of financial resources and lack of knowledge as top barriers to sustainability progress.

Also, in the Global Corporate Sustainability Report (2013), companies in the Global Compact initiative were asked to rank the top global sustainability challenges, and they cited the following:

(i)   Education (63% of respondents),
(ii) Poverty Eradication (52% of respondents),
(iii) Climate Change (52% of respondents),
(iv) Growth and Employment (49% of respondents).

We can see from the responses that companies believe that climate change and CO_2 emissions reduction is not the only sustainability issue today. Other issues, such as business growth and employment issues, were also highlighted as critical global challenges by the respondents. The employment issue was seen as having a significant effect on human dignity and self-esteem which would appear to concur with the UN Millennium Declaration (2000) which laid the groundwork for an unparalleled global effort to advance the principles of human dignity, equality, and equity everywhere. More recently, Panzaru and Dragomir (2012) also declare that business must be aware of the components that constitute sustainable practices as it is thought that their activities influence local, national, and global level. Concerning the Global Corporate Sustainability Report (2013), it should be noted that the response rate to the anonymous online survey (primary methodology) was just 25%. Questions could be asked of the commitment of the 75% of businesses who did not respond, and indeed the validity of this much-quoted report could be questioned.

3.2.2 The Economic Aspect of Sustainability

Economist Milton Friedman (1970) famously said that a business has one and only one social responsibility, namely, 'to use its resources and engage in activities designed to increase its profits'. Shaw (1988) supports Friedman and claims that he was wrongly criticised in subsequent literature and agrees that it is not the role of businesses to solve social problems such as inflation, unemployment, and pollution.

These problems demand national consensus solutions and those solutions should be hammered out by publicly elected and accountable officials. In the moral/spiritual sphere, the Roman Catholic Church appears to agree with Shaw (1988). In Deus Caritas Est (2006) it is claimed that the formation of just structures concerning the state and society is not directly the duty of the Catholic Church, but belongs to the world of politics, the sphere of the autonomous use of reason. However, the Catholic Church does claim to have an indirect duty to contribute to the purification of reason and to the reawakening of those moral forces without which just structures are neither established nor prove useful in the long run (Benedict XVI 2006). The two spheres of rite and reason are uniquely distinct, yet always interrelated. Mea and Sims (2019) argue that in the economic sphere, people are not just one more element in a means of production; they, humans, *are* the purpose. So, from this perspective, therefore, it can be claimed that faith and moral reasoning may indeed become a foundation upon which businesses and people can grow and become economically strong mutually. Wealth, and the generation of wealth, in itself is not contrary to Catholic social teaching; it is an obligatory dynamic (Mea and Sims 2019).

Entrepreneurship in business is also identified in the literature as an essential component of economic sustainability (Sautet 2013). One of the subsets of entrepreneurship is the term 'social entrepreneurship', which means people engaging in a business activity where the higher priority is given to promoting social values and development rather than capturing economic value. Social entrepreneurship is the topic of a working paper by Mair and Marti (2005), where the authors claim that the concept of social entrepreneurship is poorly defined and its boundaries to other fields of study are still fuzzy. The authors argue that it is difficult if not impossible, to quantify socioeconomic, environmental and social effects and significant efforts are needed to correct this deficiency. Indeed, in a discussion paper on social entrepreneurship highlighting ethical issues, Zahra et al. (2009) suggest that the motives of some social entrepreneurs are questionable, and they often apply new and untested organisational models, which raises concerns about the accountability of the individuals involved. They claim that social entrepreneurs share many of the same characteristics as their for-profit cohorts, i.e. *risk-taking*, *proactiveness*, and *independence* and that egoism can drive them to follow unethical practices (Zahra et al. 2009).

3.2.3   The Environmental Aspect of Sustainability

Many companies nowadays seek to protect the natural environment by making their energy usage more effective and efficient as Ireland attempts to reach national government targets that must be met concerning greenhouse gas emission levels. It is argued by many scientists that human activity is contributing to the phenomenon of *global warming* (Intergovernmental Panel on Climate Change 2013). It is claimed that by reducing our dependence on burning fossil fuels in traditional electricity generation plants and increasing our renewable energy options, we are reducing the amount of CO_2 emitted into the atmosphere. It is hoped that this strategy helps to slow down the phenomenon of global warming and protect the natural environment (Baqer 2011). As a result of higher energy awareness and much publicity of the alternative energy options, many companies are including Corporate Social Responsibility (CSR) as a strategic issue within their organisations. A significant strand to CSR appears to be measuring and reducing the carbon footprint of the business due to its energy consumption. However, a case study by Kealy (2014) suggested that the benefits of installing alternative energy systems may not be as effective as promised and requires further investigation. Sweeney (2009) carried out a study of CSR activities in a range of large, and SME's in Ireland. The author concluded by claiming that companies who take their social responsibility seriously have many advantages, among which are as follows:

(i)   Strong social reputation,
(ii) Good employee attraction,
(iii) Positive motivation and employee retention,
(iv) Easier to attract consumers,
(v)   Increased loyalty among customers.

This research found that the main barrier to CSR activity by SME's operating in Ireland is *time*, followed by *cost*, and *lack of human resources*. It should be noted that this study was carried out at the start of the downturn in the Irish economy, and many of the findings may have reflected extreme business operating conditions at that time.

3.2.4 The Human Aspect of Sustainability

The idea of human 'needs' figure prominently in the Bruntland (1987) definition of sustainability. The need for people to have the opportunity to work has long been established in the literature (Gersick et al. 2000). It is claimed that employees are more attached to their jobs with firms that engage in high levels of CSR (Carnahan et al. 2015; Ali and Jung 2017). Sison and Fontrodona (2013) claim that work is not a mere commodity or factor of production, but that it is an important opportunity for the person to develop, not just their craftsmanship, but also their moral and intellectual virtues. When the needs of each employee are met, it helps to contribute to the common good of the firm (El Akremi et al. 2018). The employees are 'stakeholders', of the company, as are shareholders, customers, suppliers, competitors, the government, and the community. In comparison with 'shareholder' theory (or a purely financial theory of the firm), Sison and Fontrodona (2013) claim that 'stakeholder' theory presents a broader and more realistic view of the corporation as a socially embedded institution. Sandelands and Hoffman (2008) in a working paper, argue that businesses progression towards a sustainable future has been limited to date. Sustainability is a well intentioned and burgeoning movement, but the movement lacks the resolve and authority needed to bring the economy, the environment and society into sustainable relations. The authors claim that the reason for this lack of progress is that people focus too narrowly on what academics, policymakers and business executives can do with market forces and government regulations to promote sustainable development, and in doing so, we concede too much to private interests. Sandelands and Hoffman (2008) do not condemn private interests in the marketplace, but they argue that for a real idea of sustainability to be conceived or achieved, economic reasoning and governmental policy must be inspired and informed by faith. The authors in this working paper are strongly influenced by their Christian faith, and therefore some researchers may disagree with a faith input to such a discussion. Sustainable development influenced by a spiritual attribute is also expressed by Tucker and Grim (2007) in an existential reflection piece. Note that Mary Evelyn Tucker and John Grim are the directors of the Forum on Religion and Ecology at Yale University in the USA. In addition to the concept of human needs, some scholars have sought to examine the link between business and human rights. Sen (2004), based on a strong philosophical foundation, argues that human rights are seen to derive from the inherent dignity of the human person. The concept of a human 'right' belongs to a branch of morality which is specifically concerned to determine when one person's freedom, or right, might be limited by another's. Many acts of law have been inspired by a belief in some pre-existing rights of all human beings. However, Sen (2004) questioned whether the law is the pre-eminent or even a necessary route through which human rights can be pursued. To encourage business leaders to adopt an ethos of human-dignity centred management, Mea and Sims (2019) propose a practical framework to support just and effective managerial practices. The framework is highly influenced by Catholic Social Doctrine and Teaching (CSD/T), and the authors state that some business leaders may not be open to accepting religious premises.

3.2.5 Measuring Sustainable Development Practices

Even if businesses are operating ethically, it appears from the literature that there is a difficulty in measuring sustainable development in their organisations (Mair and Marti 2005) with some researchers constructing sustainable development measurement frameworks based on natural resource inputs and human welfare outputs (Yan et al. 2018). A comment often made concerning management theory is 'if you cannot measure it, you cannot manage it'. Elkington (1997) did attempt to propose a new framework to measure sustainability in American corporations. The concept was called the 'Triple-Bottom-Line' (TBL) accounting framework and undertook to measure not just the traditional measures of profit, but also included environmental and social dimensions as well. The three interrelated aspects are sometimes referred to as the 3P's, Profit, People, and Planet. Recent research (Staniskiene and Stankeviciute 2018) claim that while the environmental (Planet) and economic (Profit) dimensions could be evaluated using quantitative indicators, the social (People) aspect requires a balance between quantitative and qualitative indicators. Jamaludin et al. (2018) developed a sustainability index that is used to measure sustainability performance with a particular focus on palm oil millers. This measuring technique then allows millers to differentiate between the performances of several mills. Elkington (1994) considered some of how businesses are developing what he terms as 'win-win-win' strategies to simultaneously benefit the company, its customers, and the environment. Subsequent critique of the work of Elkington (1997) was conducted by Milne and Gray (2013) who claimed the TBL language of businesses' management, measurement, and reporting processes of the three elements is out-dated and belonged to the late 1990s and early 2000s. The authors questioned the TBL concept and its ability to contribute to a sustainable future, even suggesting that it may be acting against the sustainable development of organisations. Milne and Gray (2013), however, do not provide an alternative model that allows a business to incorporate its economic, environmental, and social performance into its management and reporting processes. The investors who invested €280,000 in a wind turbine project (Kealy 2017a, b) while they may agree with Milne and Gray (2013) and be discouraged with the outcomes of their TBL efforts (CO_2 emission reduction and cost savings) at a minimum they had access to measured environmental and economic information by which to evaluate the wind turbine project. Therefore, independent studies such as the studies by Kealy (2017a, b) have the potential to contribute to the debate on TBL issues where the TBL concept may still be valid, but accurate measurement appears to be a key component.

The motives for businesses adopting sustainable practices having been questioned, as history has taught us that not all business activity appears to have been used as a tool to help to bring equity to the world. The Irish banking industry in the first decade of the third millennium is a case in point (Whelan 2014). However, despite the public dissatisfaction with the banking organisations in subsequent years, there was no one to shout 'stop' during the years of unbridled economic growth where many residents benefited from the public policies and bank-funded activities of private investors. An article by Peloza et al. (2012) discusses how getting a high rating from a global sustainability index might be used solely to give the firm a competitive advantage. It should be noted that the article by Peloza et al. (2012) is limited in that it only commented on large American Multinational firms like Walmart, Disney, Ford, and IBM among others and concludes that the public has a homogenous perception of the sustainability efforts in the firms. Corporations have problems positioning themselves differently in the environmental arena from their competitors. They all look the same. An example of a global sustainability initiative is the Global Reporting Initiative (GRI), and this is intended to provide a generally accepted framework for reporting on an organisations economic, environmental, and social performance. Commenting on this initiative, Milne and Gray (2013) argue that the GRI provides conditions that reinforce business-as-usual and higher levels of un-sustainability. The GRI is an independent institution, and some of the most influential critics of its development have been its apparent reluctance to define sustainability and sustainable development (Wackernagel 2002). Perhaps some companies perceive the 'sustainability logo' from a marketing perspective (described in Sect. 2.2.1.1) instead of concrete efforts in attempting to align their business strategy with their sustainability strategy (described in Sect. 2.2.1.3).

In conclusion, therefore, there is much written regarding sustainability development initiatives and many pieces of research in this area. It should be noted that many publications are American-based, with minimal Irish-based research. It appears that critical components of economic, environmental, and human aspects of sustainable business development have drawn much discussion and investigation. While the TBL concept has its critics, it may be that what is needed to develop the TBL framework is an enhancement of the *measurement* dimension of the three critical components of sustainability reporting.

3.3   Research Methods/Methodology

The area of interest for this chapter is the sustainability attitudes and perceptions within a broad range of Irish organisations currently. The chapter has an ethnographic emphasis as it probes people's thoughts, theories, and points-of-view on sustainable development in their real-life environments (Moger and Bagley 2019). The survey methodology is employed, and data is collected utilising the commonly-used interview technique (Choumert-Nkolo et al. 2019). The method of analysis on the qualitative data is thematic analysis on the interview data. Thematic analysis is a means of analysing qualitative data in a rigorous and methodical manner (Nowell et al. 2017). The qualitative data were obtained from interviews which consisted of the researcher asking eleven pre-determined questions (see 'Appendices for Chapter 3' for a list of questions) and taking notes of the responses. Thematic analysis was undertaken by developing codes in the analysis of the interviews. It was decided to interview a wide range of businesses operating in Ireland. As the initial business data collection/analysis developed, it became clear that it would be beneficial to have input from other non-business but influential organisations (GAA, IFA, Roman Catholic Church, Trade Unions, Charity organisations) because it was considered that they contribute to the debate on sustainable business development, which appears to have an influence on financial, environmental, and human spheres. Themes derived from the research data were developed based on the rich data obtained as a result of the wide-ranging organisations who took part in the project. Each participant in the research project held a middle-management or top-management role within their organisations and included the following, with initials used to mask their identity:

(i)   EC, Senior Manager, Department of Communications, Energy, and Natural Resources (DCENR),
(ii)   TE, Franchise Owner, Spar Supermarket, Dublin,
(iii)   NO'C, Director, Think-tank for Action on Social Change, TASC (Not-for-profit Limited Company), Dublin 2,
(iv)   EB, Facilities Officer, Croke Park (Gaelic Athletic Association),
(v)   EM, Senior Personnel, Roman Catholic Church,
(vi)   KK, Head of Engineering Department, Dublin Institute of Technology,
(vii)   GO, General Manager, Geith International, Meath,

(viii) KM, Programmes Policy Officer, Gorta, NGO Charity Organisation,
 (ix)   CB, Corporate Social Responsibility and Funding Officer, Stobart Group (PLC), UK,
  (x)   ED, Senior position, Irish Farmers Association,
 (xi)   DB, Senior position, Irish Congress of Trade Unions (ICTU),
(xii)   PL, Duty Manager, Buswells Hotel, Dublin,
(xiii) MH, Partner, Atlantic Bridge (Ventures), Dublin.

3.3.1 Ethical Issues

Many of the participants in this study are well-known members of Irish business/organisations/society; their names instantly recognisable. It should be noted while agreeing to take part in this research, the participants specifically requested that the researcher state in his completed work that the opinions expressed by each respondent were their personal views and may not reflect the official view of their organisations. The researcher assured the participants that this fact would be made clear in the research.

3.3.2 Validity and Reliability

Every effort was made to reduce bias in this study. The questions were reviewed by some experts and were constructed in such a manner as to motivate the respondents to answer as completely and honestly as possible.

3.3.3 Data Analysis

Data analysis was conducted using the thematic analysis approach (Braun and Clarke 2006). Firstly, it was necessary for the researcher to become much familiarised with the data. This stage was helped by the researcher transcribing the verbal data into written form on the researchers PC. Secondly, initial codes were developed. As a result of the coding stage, initial themes were developed, which were subsequently reviewed. The next step in the thematic analysis method was to define and name the themes that emerged. The final stage was to produce a model based on the qualitative data analysis process.

Coding allowed for the simplification and focused on pertinent characteristics of the data. Labels were attached to the relevant features (codes). The purpose of the theme development is to interrogate the meaning of the text and identify patterns that underlie the themes. Identified themes link substantial portions of the data together

(McIntosh et al. 2013). Reviewing of the themes assesses whether there is enough, focused, data to support the themes. The remaining themes were subsequently named. Direct quotes from the participants were included in the study.

During the manual coding process, the researcher regularly revisited the data clarifying what was being said and which code it may indicate. This painstaking task was documented continuously during this process with a memo writing where ideas were jotted down fixedly but in a dynamic way that they could be revisited and altered as the process progressed. The initial coding process allowed for the identification of 33 codes. These codes can be viewed in 'Appendices for Chapter 3'. The first twelve codes in the list enabled the researcher to develop a theme, namely, the 'Varied Perceptions of the Concept of Sustainability' theme. A review of this first overall theme was undertaken to allow four sub-themes to be named (Social equality—Energy—Food—Ethics/Morals). Interpretation of the data codes ensured that five main themes developed from the qualitative data. These themes included (i) the varied perceptions of the concept of sustainability (and its sub-themes of social equality/energy/food/ethics/morals) (ii) regulation compliance (iii) organisational leadership in sustainability (iv) marketing (v) human aspects. Indeed a core theme in this research appeared to develop, that being the varied subjective nature among the participants as what organisational sustainability meant to them.

3.4 Research Findings

From the outset, it was clear that specific codes and themes were developing from the manual analysis of the data. These were narrowed down to five main themes. These included the following:

(i)   Varied Perceptions of the Concept of Sustainability,
(ii)   Regulation Compliance,
(iii) Organisational Leadership in Sustainability,
(iv)   Marketing,
(v)   Human Aspects.

3.4.1 Varied Perceptions of the Concept of Sustainability

(*Italics indicate direct quotes from the interviewees*)

There appears to be a wide-ranging understanding of the concept of sustainability in organisations. Because of the wide-ranging use of the term, it was necessary to pinpoint the focus of the research questions to each participant organisation at the outset. This core theme links the four sub-themes of social equality, energy, food, and ethics/morals.

Sustainability (Social Equality) One of the participant organisations very aware of the societal equality connotation of the sustainability term was the not-for-profit organisation, TASC, which claimed to operate as an 'independent, progressive think-tank dedicated to promoting equality, democracy, and sustainability in Ireland through evidence-based policy recommendations'. Their Director, NO'C, declared '*we have a mission to bring about a more equal society, full employment, pay decent wages and have a good quality of life in the workplace. We are a sustainable organisation but very importantly, the charitable mission must be sustainable as well. Over the period of our eleven-year existence, we have seen our reputation increase, we have received more media exposure, and we are being invited to policy-making meetings*'. This organisation was aware that they must have a sound financial structure, but it appears to be very mission-driven with employees cognizant of the human and environmental aspects as well. Indeed, when the interview took place, concern was expressed that the external funds, mostly provided by an American-based philanthropist, were expected to reduce over the next couple of years and they hope to continue with their mission on a tighter budget. All organisations in this research appear to be operating on a stricter budget because of the challenging economic times, but the findings showed that there was unanimous agreement that businesses must be profitable to be sustainable.

Sustainability (Energy) Many of the interviewees perceived sustainability as dealing mainly with energy efficiency within their organisations. Climate change and global warming were among the two most frequent answers given when asked about the current most pressing global challenges. Both PL, Duty Manager with Buswells Hotel and TE, Franchise owner of a chain of Spar supermarkets disclosed how they '*have colour-coded bins for the different types of waste and any new refurbishment considers using Light Emitting Diodes (LED's) as the light sources*' when asked about the importance of sustainability to the management of their businesses. EB, facilities manager in Croke Park, acknowledged that '*the huge cost of utilities has also contributed to us looking at sustainability issues. We have installed a Building Management System (BMS) to help assist us with this*'. When asked about sustainability as a strategic issue within his franchise business, TE (Spar Supermarket) stated that he '*has four kids now, so I see the need to reuse and recycle*'. Interestingly, no interviewee specifically mentioned renewable energy technologies as part of their sustainable initiatives. The majority of companies described how they are 'switching off the lights' when speaking about sustainable practices, but none of the interviewees/businesses appears to have undertaken renewable energy projects, for example, wind turbines or other alternative energy resources. Only one business who took part in this research had a dedicated CSR officer in place, namely, the Stobart Group. The company, with 6000 drivers, is the most significant business to take part in this data collection process and is a UK-based logistics organisation with a branch in Ireland. The CSR Officer with the Stobart Group, CB, was evident on her companies' perception of sustainability and declared that '*sustainability is an important aspect of the running of their business but not as much at Board level as I would like. One way of increasing the exposure at Board level is to put Key Performance Indicators (KPI)* *into their sustainable management returns'*. She added that '*sustainability measures are evaluated and/or measured by the amount of money we save by using less fuel as a result of inserting Global Positioning Systems (GPS) in our trucks. This allows us to measure the "empty miles" and truck "reloads" in real-time, making for a more efficient logistics operation'*.

Sustainability (Food) Food and food production was another frequent answer when the interviewees were asked about current global challenges. At the time of the data collection, December 2013, there was a 'price war' among the large grocery chains in Ireland and vegetables were being offered at a low price. Commenting on his understanding of sustainability within his organisation, ED (IFA) declared that one of the significant global challenges to sustainability was to '*produce enough food to feed the seven billion people that exists on the planet'*. He maintained that '*this is possible, but to do it, we must manage it right. This means things like cutting out waste, but importantly, we can't be selling vegetables for five-cent. This is not sustainable'*. Considering hunger and food on a global level, Gorta the NGO charity organisation established in 1965 who works with local partners in developing countries, assists indigenous people in the move from subsistence to entrepreneurship. Their programmes policy officer, KM, stated that '*they didn't want to develop situations where the people could just survive, i.e. not hungry, but they want a bit more than that. They, Gorta, want people to have enough income to develop their (small, local) business in a sustainable way, i.e. it will survive into the future'*. KM explained that one of the main components of Gorta's strategy is to engage with Irish 'for-profit' partners, such as '*Glanbia, Intel, and the Project Management group. These organisations fund-raise for us but also provide qualified personnel to go out and set up projects'*. When asked about the most pressing global challenges, KM stated that '*food and the production of food is a challenge. We can create price spikes by treating food as a commodity. It becomes a problem when we treat food as a commodity. This will affect, for example, rice farmers'*. It is clear that for NGOs like Gorta, a significant component of sustainability means the availability of sufficient food to feed the population.

Sustainability (Ethics/Morals) All of the participant organisations were clear that ethical business practices were of the utmost importance and '*a given*' when discussing how businesses can be developed sustainably. Trade unionists DB claimed that '*we are all inter-dependent. Society must be ethical to make progress. I think that you have to conduct your affairs within the ethical boundaries'*. During the course of the interviews, the Irish banking crisis was used as an example of what can happen when ethical business procedures are not followed. KK, Engineering Department in DIT, spoke about morals within businesses when he claimed that '*An ethical business is a moral business. A business that is ethical will be sustainable, not just in the energy sense, but also in a business sense'*. The moral aspect of sustainable development is encouraged in National Schools, under the patronage of the Catholic Church, as explained by senior personnel EM who stated '*As a Church, we are constantly raising moral awareness of sustainability issues. Catholic schools are pioneers for green flags for energy use'*. Indeed, education also featured prominently when the interviewees were asked about the current sustainability challenges faced on a global level.

3.4.2 Regulation Compliance

From this research, compliance with Irish and European regulation appears to be a significant driver concerning organisational sustainability. Many of the interviewees referred to government legislation on sustainability practice and the need to adhere to it. EC, a senior manager in the Department of Communication, Energy, and Natural Resources (DCENR), argued that *'sustainability is extremely important for his department as we have overall control of energy efficiency policy, which is becoming more and more European policy. We will be judged on how our policies affect the citizens of this country. For example, it is not fair that someone living in Kerry does not have the same access to broadband as someone living in Dublin, just because of location'*. The government was seen to have an essential role in organising and encouraging sustainable practices within businesses who took part in this research. For example, when trade unionist DB was asked about social entrepreneurship, he replied *'it is not just a problem for businesses to solve. Government must be very involved. Charities are like individual building bricks, but the government must put all the bricks together. The state has a big responsibility to get an ordered society'*. This sentiment about the governments' role concurs with the views expressed by GO, General Manager in Geith International. From a UK perspective, CB (Stobart Group) stated that *'we have a clear target to reduce CO_2 emissions in line with the UK directive. There are bigger problems than us as logistics provider, but we contribute to the problem so must play our part as well'*. The Stobart Group is not aligned to any sustainability index as *'we see that as for the global brands and we are not that big yet'*.

KK, Head of an engineering department in Dublin Institute of Technology, cited *'energy costs and government policy'* as the main drivers of the sustainability issues within his institute. This drive influences the processes within the institute, as well as the programmes offered by the institute. ED (IFA) maintained that an external factor that caused its members to consider sustainability is *'legislation in relation to, for example, nitrate usage and water quality'*. On the positive side, he also acknowledged that the quality of water in rural Ireland is improving. The farmers' legislation appears to be significantly influenced by the European central government, which provides significant funds to the Irish farming community. Water and water quality was expressed by many interviewees regarding current sustainable global challenges. When asked about external factors that influence the company looking at their heavy plant construction business, GO, General Manager with the Multi-National Corporation, Geith International explained *'the change in legislation affected the way we develop our products. We need to adapt to the new situation. So, there are external factors, if the government change the laws, we must follow'*. He also declared that their business must take steps to ensure that their business practices do not contaminate the river, running beside their factory. He made the point *'that our factory is in area surrounded by nature, fields, rivers etc. so we must pay special attention to protecting that. We must not decimate the local environment'*.

3.4.3 Organisational Leadership in Sustainability

Leadership on sustainability issues emerged as a theme among the participants of this research project, and all the participants agreed that sustainability is a strategic issue. EB, facilities manager in Croke Park, explained that *'we are seen as the flag-ship of the organisation'* so are seen as having a leadership role in the overall running of the Gaelic Athletic Association (GAA). A similar view is expressed by KM, programmes policy office with Gorta, who argued that *'leadership is very important in management. It inspires people to see somebody who are passionate about sustainable development'*. Again, MH, Partner in Atlantic Bridge (Ventures), declared *'a top sustainability challenge is developing management teams. It is very important to have good leadership, this is central to sustainability'*. DB (ICTU) also identified that *'leadership is very important. Leadership is perhaps someone not having great ability, but being able to recognise great ability in other people'*. He maintained that social entrepreneurs could help to make a difference in disadvantaged areas, but *'societal problems are not just for businesses to solve, government must be very involved'*. In the education sector, KK declared that *'we must be seen to be a leader in the sustainability area. Programmes and courses have been developed to progress sustainability issues. We have made capital investments in equipment and laboratories to reflect these new sustainable technologies. For example, a significant sum of money has been invested in a lighting laboratory and also we have invested in the siting of Photo-Voltaic (PV) panels on the college roof'*.

3.4.4 Marketing

Marketing issues were identified as a theme for some, though not all, of the interviewees. From some participants, it was clear that sustainability accreditation could be used as a competitive advantage and promoted in the brand-building efforts of the business. CB of the Stobart Group acknowledged that *'as we are a Public Limited Company (PLC), investors are more interested if we are seen to be operating in a responsible way'*. PL of Buswells Hotel disclosed *'we are aligned with the green hospitality ethos. This sets targets on items like waste and energy usage. This helps us with the marketing of the business, and we have a sticker at the entrance to the premises showing the green hospitality accreditation'*. The green marketing approach has also been adopted by the GAA in their headquarters at Croke Park. EB, facilities manager, explained that when corporations are holding an event in Croke Park *'we let people know about our sustainability achievements and encourage them to follow our sustainable practices when attending'*. They hope that this ethos filters down through clubs and organisations right throughout the country.

3.4.5 Human Aspects

The majority of the participants in the study had strong views on the importance of the human aspect when discussing sustainability. CB of the Stobart Group pronounced that *'people are what make us. Our company have a very good training programme in place. We have a three-tier system where office employees have an opportunity to develop through the stages and eventually attend a leadership course. We are currently revising our charity policy to make our employees more empowered. Each employee gets a certain amount to use in their local charity'*. Empowering employees is also deemed to be essential for PL, duty manager in the hospitality industry, and when asked about the personal development of their employees, revealed that *'we have a number of employees here for up on forty years. This is unusual in our type of business as there is generally a big turnover in staff. We give each employee an opportunity to work, and move, to different sections of our business. We also treat each employee with respect. The management team is very hands-on and we encourage a good team ethic'*. Team ethic in a busy, city-centre, store is deemed essential for TE, franchise owner of a chain of five Spar grocery outlets, and he expressed the view that *'the most important part of my business are the human beings that I have working with me. It is a people business and I have some employees with me for a long time. I treat them with good manners like human beings, give them a good safe environment, and help them to develop as a person. When I am employing someone, the last thing I look at is what is on the CV. I just have a conversation, and I know if that person can work with me or not. By paying my staff a fair wage, I believe that I am contributing to society'*.

EC, senior manager in the DCENR, professed that *'we see personal development as part of our sustainability efforts, but we also have personal responsibility. We have a responsibility to, for example, switch off unused lights and boil only a small amount of water in the kettle when we only need a small amount. Very importantly, we have taken on four unemployed people in the past couple of years from the Job-Bridge programme. These were highly qualified people and it worked really well. All of the participants have gone on to either full-time positions or further education. It has been a positive experience for both our department and the participants'*. The importance of treating people with respect is also discussed by MH, a Partner in a Venture Capitalist business, who acknowledged that *'nowadays people are more educated, they are more conscious of the environment, and of fairness. People must be treated with respect. Our company has lots of high-level graduates who want better. Fifty years ago, people were not aware of these issues like they are now'*. It seems that in recent years, more emphasis is put on structures to encourage the development of employees within businesses as described by GO, General Manager with Geith International, who explained *'we have development plans and the most important aspect is the people aspect. You can have a great product and a great market but without the right people, it is dead. One of the key factors of success is people development; we need to invest in our people. We don't just send them on external courses, but we structure their development needs. If there are weaknesses we don't see that as a negative, we see it as a potential for improvement. Some people respond best to coaching and mentoring. We empower them and this provides a challenge to make their own decisions. If they make a mistake, we don't look down on them, because a mistake is never done on purpose. The main thing is that we can learn from the mistake, we give them opportunities to do that'*.

EB, Croke Park, viewed the concept of social entrepreneurship as being central to the success and longevity of the GAA. They are in existence for over 125 years, and *'local people have taken a chance with the organisation in their community and did not get any economic reward. The fruits have been seen in how we have expanded, not just locally in every parish in Ireland, but on a world-wide stage like England, America, and Australia'*. Many of the people involved in the GAA are also farmers and the social link with other people in similar situations as themselves contributes to their well-being. ED (IFA) declared that many of their members attend Open Days and this is important as *'farmers tend to work a lot in isolation so the opportunity to meet people from rural Ireland is a positive thing. It encourages camaraderie among farmers'*. KM (Gorta) spoke about the importance of human aspects of business sustainability. She identified the importance of keeping people happy within their work organisation *'if people leave the organisation, they take a lot of corporate knowledge with them. This is not good for sustainable development'*.

3.5   Research Analysis

This was a challenging piece of research to conduct, not least because of the wide-ranging views and opinions from the broad cross-section of business and non-business organisations. It would appear from the results that if sustainability was a stand-alone brand, from a marketing point of view, it needs much work on its positioning strategy (Tyagi and Raju 2018). Because of the overuse of the sustainability term (brand), the target markets are not well-defined. It was clear from all the participants that sustainable development ranged from reducing their CO_2 emissions by becoming more energy-efficient, to maintaining profits, to having access to a good education, to having access to good quality water, to having enough food on the table and that it was produced in an ethical and moral fashion. The instant that the term sustainability was mentioned, there appeared to be an absence of a clear picture of what the researcher meant. This lack of a firm view is in complete contrast to what we are told that brand-building should be (Tyagi and Raju 2018), and may even lead to adverse effects on the business marketing efforts. This research has demonstrated that there was no single concise understanding of sustainable development among the participants; the topic seemed vastly subjective.

3.5.1 Organisational Sustainability

When contact was made with each of the participant organisations for this research, it was interesting to note that it was generally the facilities manager or the CSR manager to whom the query was directed. It was not seen as the remit of the CEO, or the Board of Directors to participate in this research process. This finding possibly highlights a problem with the general area of sustainability and the specific area of sustainable business development. It would appear for many organisations that their business sustainability policies are governed by one individual under the corporate social responsibility folder rather than an all-encompassing mission of the business. It could be argued that senior management must include sustainability at a strategic level within the company with every stakeholder having a contribution to sustainable development issues of the business, this would consist of employees of the company, shareholders, customers, suppliers, competitors, the government, and the community in which the business operates (Sison and Fontrodona 2013).

3.5.2 Interconnection

In this piece of work, certain critical elements of sustainability appeared very important to the research study. It was noticed by the researcher that these elements were a common thread running through the responses of the participants. These are also the three elements identified by Elkington (1994) in his triple-bottom-line theory:

(i)   Profit,
(ii) People,
(iii) Planet.

Most of the respondents highlighted these components as essential aspects of any business sustainability model. The inter-dependence and synergies that exist between the three main components concerning sustainability are best summarised by way of a Venn diagram (Fig. 3.1) and graphical representation (Table 3.1).

The three elements of Fig. 3.1 (Person, Planet, and Profit) are interrelated. For example, if the business is making a reasonable profit, then the Person working in the business also benefits because the company can afford to pay a fair wage, as seen in Table 3.1 (Blue column). Similarly, if the Person working in the business does not allow any contaminants to flow into a river, as stated by GO the General Manager in Geith International, then the planet benefits because there is clean water as seen in Table 3.1 (Green column). Indeed, GO General Manager, demonstrated strong leadership in all three areas and appeared to epitomise important activity in the central overlapping area of the Venn diagram. This central area seems to encompass all three well-established elements of profit, planet, and person. All respondents highlighted the importance of environmental issues, and social responsibilities as well as improving profit margins of businesses, albeit at times in aspirational or

Fig. 3.1 Three P's of
CSR/SD

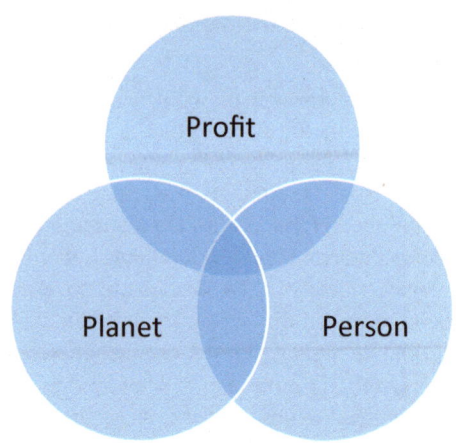

Table 3.1 Synergies between Profit/Person (blue), Profit/Planet (orange), and Person/Planet (green)
in sustainable business development

Profit/Person	Profit/Planet	Person/Planet
Fair Wage	Reduce Energy Usage	Clean Water
Promotion Opportunities	Invest in Renewable Energy sources	Fresh Air
Enough Food to feed all people	Corporate Social Responsibility	Environmental Ecology
Provide for Peoples Material, Emotional, and Spiritual needs	Proper Waste Management	Preserve Natural Habitats
Afford people the opportunity to work (locally)	Carbon Capture and Storage	Engaging Employees
Safe Working Environment	Implement Correct End-Of-Life Procedures for Products	Managing Environmental Risks
Secure Pension	Ethical Practises	
Provide for Children's Education		

Profit/Planet/People
Sustainability

haphazard fashion. The data also pointed to the interconnectedness between peoples on our planet. According to some of the respondents, our activities in one part of the world seemed to affect other people, perhaps living at the other end of the globe. This interconnection between profit, people, and planet concur with writings from Shaw (1988), Elkington (1994), and Benedict XVI (2009).

There was little evidence found in this research as to the desire of Irish businesses to be part of any global sustainability indices. None of the companies who took part in this research was part of the United Nations Global Compact. However, many of these businesses were accredited to the national, Irish standards, for example, 'greenhospitality', 'ISO 50001' (Energy Management Systems), 'ISO 14001' (Environmental Management Systems), and 'ISO 20121' (Sustainable Events Management). These standards seem to be an essential aspect of their business from a leadership and marketing platform.

3.5.3 Business Sustainability Model

From the findings of this study, it is clear that for businesses to be developed sustainably, some specific essential input components should be considered by organisational decision-makers. The four fundamental tenets are identified (Fig. 3.2) as follows:

(i) **For-Profit Ethos**. Each interviewee agreed that an organisation needs to be profitable to be sustainable. A sustainability accreditation can be used as a competitive advantage (Sect. 3.4.4) and promoted in the brand-building efforts of the organisation. A focus on profit was not seen as adverse to a focus on sustainability.

(ii) **Not-For-Profit Ethos**. All of the NGO/Charity groups interviewed were very much focused on a mission element and were very aware of issues other than

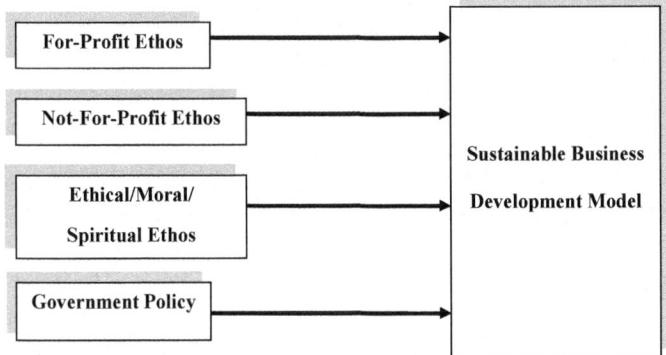

Fig. 3.2 Essential input components of sustainable business development model

financial matters, for example, social and environmental issues. A profitable business has the potential to contribute positively to the society in which they operate.

(iii) **Ethical/Moral/Spiritual Ethos**. The human aspects (Sect. 3.4.5) and needs of workers, their families, and the broader social community were identified as essential considerations in business sustainability as well as the general business ethic (Sect. 3.4.1) adopted by a company or organisation. An ethical/moral/spiritual ethos is seen as a trait inherent in organisational leaders who promote and pursue social goals that extend beyond their legal responsibilities (Carroll and Shabana 2010).

(iv) **Government Policy**. This government role was also identified as influential in designing policies that enhance sustainable business development and ensuring that organisations comply with regulations (Sect. 3.4.2).

An example of an interviewee who encompasses all of the proposed inputs is GO, General Manager in a Multi-National Corporation with a subsidiary in Ireland whose operations have survived through the recent challenging economic times.

3.6   Discussion

It appears from this study that the promotion of sustainable business development practices in Ireland has suffered as a result of an overuse of the term 'sustainability' in recent years by many organisations in various situations. The word appears to be somewhat ambiguous and subjective. Ironically, companies market themselves as providing sustainable products and services, while this research has shown that the sustainable brand is not clearly defined in peoples' minds. Perhaps it is time to determine a new name for the concept or introduce an objective component into the Sustainable Development/Corporate Social Responsibility sector.

The United Nations Global Compact sustainability reporting initiative appears to have had little influence in Ireland concerning the adoption of sustainability processes. This researcher finds himself concurring with opinions expressed by Benedict XV1 when commenting on globalisation in his Caritas. In Veritate encyclical letter declared that what is needed is a '*reform of the United Nations Organisation, and likewise of economic institutions and international finance so that the concept of the family of nations can acquire real teeth*'.

From this piece of qualitative research using the thematic data analysis method, a new sustainable development theory has been established, signifying four essential components to be considered by organisational management in their decision-making process. The vital components are as follows:

(i)   For-Profit Ethos,
(ii) Not-For-Profit Ethos,
(iii) Ethical/Moral/Spiritual Ethos,
(iv) Government Policy.

3.7 Conclusion

This was a challenging ethnographic study primarily because of the ambiguity and subjectivity surrounding the concept of sustainable organisational development. Although most of the survey participants were unambiguous in their understanding of sustainable development within their organisations, the vast and varied attitudes and perspectives of 'sustainable development' made for complex analysis in this project. The idea of sustainability appears to be 'nebulous and contested' as claimed by Sandelands and Hoffman (2008). The importance of strong leadership in business is borne out by the findings in this research (Sect. 3.4.3). The study also found that for organisations to develop sustainably, they require leaders with a clear strategic vision who are guided by a strong ethical/moral compass. Sustainability is achievable by incorporating the critical input components of the new proposed theoretical model into the organisational culture (Fig. 3.2).

References

Ali, M. A., & Jung, H.-J. (2017, April). CSR and the workplace attitudes of irregular employees: The case of subcontracted workers in Korea. *Business Ethics: A European Review, 26*(2), 130–146. https://doi.org/10.1111/beer.12146.

Baqer, S. M. (2011, June). Hunting for true green consumers: A multicultural investigation of consumers' genuine willingness to share the responsibility of saving the environment'. *Global Conference on Business & Finance Proceedings, 6*(2), 325–337.

Benedict XVI. (2006). *Deus Caritas Est* [God is love]. Encyclical Letter of the Supreme Pontiff Benedict XVI.

Benedict XVI. (2009). *Caritas in Veritate* [Charity in truth]. Encyclical Letter of the Supreme Pontiff Benedict XVI.

Braun, V., & Clarke, V. (2006). Using thematic analysis in psychology. *Qualitative Research in Psychology, 3*(2), 77–101.

Bruntland Commission Report. (1987). *World Commission on Environment and Development: Our common future*. New York, NY: Oxford University Press.

Carnahan, S., Kryscynski, D., & Olson, D. (2015). How corporate social responsibility reduces employee turnover. *Academy of Management Annual Meeting Proceedings, 2015*(1), 1. https://doi.org/10.5465/ambpp.2015.14792abstract.

Carroll, A. B., & Shabana, K. M. (2010). The business case for corporate social responsibility: A review of concepts, research and practice. *International Journal of Management Reviews, 12*(1), 85–105. https://doi.org/10.1111/j.1468-2370.2009.00275.x.

Choumert-Nkolo, J., Cust, H., & Taylor, C. (2019, May). Using paradata to collect better survey data: Evidence from a household survey in Tanzania. *Review of Development Economics, 23*(2), 598–618. https://doi.org/10.1111/rode.12583.

El Akremi, A., Gond, J.-P., Swaen, V., De Roeck, K., & Igalens, J. (2018, February). How do employees perceive corporate responsibility? Development and validation of a multidimensional corporate stakeholder responsibility scale. *Journal of Management, 44*(2), 619–657. https://doi.org/10.1177/0149206315569311.

Elkington, J. (1994). Towards the sustainable corporation: Win-win-win business strategies for sustainable development. *California Management Review, 36*(2), 90–100.

Elkington, J. (1997). *Cannibal with forks: The triple bottom line of 21st century business*. Gabriola Island, BC: Capstone.

Eweje, G. (2011, May/June). A shift in corporate practice? Facilitating sustainability strategy in companies. *Corporate Social Responsibility & Environmental Management, 18*(3), 125–136.

Friedman, M. (1970, September 13). The social responsibility of business is to increase its profits. *New York Times*, 122–126.

Gersick, C. J. G., Dutton, J. E., & Bartunek, J. M. (2000, December). Learning from academia: The importance of relationships in professional life. *Academy of Management Journal, 43*(6), 1026–1044.

Intergovernmental Panel on Climate Change. (2013). Available from http://www.un.org/climatechange/blog/category/ipcc/. Accessed on 24 February 2014.

Jamaludin, N. F., Hashim, H., Ab Muis, Z., Zakaria, Z. Y., Jusoh, M., Yunus, A., et al. (2018, February). A sustainability performance assessment framework for palm oil mills. *Journal of Cleaner Production, 174,* 1679–1693. https://doi.org/10.1016/j.jclepro.2017.11.028.

Kealy, T. (2014, April). Financial appraisal of a small scale wind turbine with a case study in Ireland. *Journal of Energy and Power Engineering, 8*(4), 620–627. https://doi.org/10.17265/1934-8975/2014.04.004.

Kealy, T. (2017a, February). Stakeholder outcomes in a wind turbine investment; is the Irish energy policy effective in reducing GHG emissions by promoting small-scale embedded turbines in SME's? *Renewable Energy, 101,* 1157–1168. https://doi.org/10.1016/j.renene.2016.10.007.

Kealy, T. (2017b, October). Does an embedded wind turbine reduce a company's electricity bill? Case study of a 300-kW wind turbine in Ireland. *Journal of Business Ethics, 145*(2), 417–428. https://doi.org/10.1007/s10551-015-2837-44.

Mair, J., & Marti, I. (2005, February). Social entrepreneurship research: A source of explanation, prediction, and delight. *Journal of World Business, 41*(1), 36–44.

McIntosh, K., Martinez, R. S., Ty, S. V., & McClain, M. B. (2013, June). Scientific research in school psychology: Leading researchers weigh in on past, present, and future. *Journal of School Psychology, 51*(3), 267–318. https://doi.org/10.1016/j.jsp.2013.04.003.

Mea, W. J., & Sims, R. R. (2019). Human dignity-centered business ethics: A conceptual framework for business leaders. *Journal of Business Ethics, 160*(1), 53–69. https://doi.org/10.1007/s10551-018-3929-8.

Milne, M. J., & Gray, R. (2013). W(h)ither ecology? The triple bottom line, the global reporting initiative, and corporate sustainability reporting. *The Journal of Business Ethics, 118,* 13–29.

Moger, P., & Bagley, C. (2019). A space for policy legacy: An ethnographic exploration of a secondary school's commitment to creativity after national policy priorities have changed. *Ethnography and Education, 14*(1), 101–118. https://doi.org/10.1080/17457823.2917.1396544.

Nowell, L. S., Norris, J. M., White, D. E., & Moules, N. J. (2017). Thematic analysis: Striving to meet the trustworthiness criteria. *International Journal of Qualitative Methods, 16,* 1–13. https://doi.org/10.1177/1609406917733847.

Panzaru, S., & Dragomir, C. (2012, December). The considerations of the sustainable development and eco-development in national and zonal context. *Review of International Comparative Management, 13*(5), 823–831.

Peloza, J., Loock, M., Cerruti, J., & Muyot, M. (2012, Fall). Sustainability: How stakeholder perceptions differ from corporate reality. *California Management Review, 55*(1), 74–97.

Sandelands, L. E., & Hoffman, A. J. (2008). Sustainability, Faith, and the Market. *Worldviews: Global Regions, Culture & Ecology, 12*(2/3), 129–145. https://doi.org/10.1163/156853508X359949.

Sautet, F. (2013, March). Local and systemic entrepreneurship: Solving the puzzle of entrepreneurship and economic development. *Entrepreneurship: Theory and Practice, 37*(2), 387–402.

Sen, A. (2004). Elements of a theory of human rights. *Philosophy & Public Affairs, 32*(4), 315–356.

Shaw, B. (1988). A reply to Thomas Mulligan's "critique of Milton Friedman's essay, 'The social responsibility of business is to increase Its profits'". *Journal of Business Ethics, 7,* 537–543.

Sison, A. J. G., & Fontrodona, J. (2013). Participating in the common good of the firm. *Journal of Business Ethics, 113,* 611–625.

Staniskiene, E., & Stankeviciute, Z. (2018, July). Social sustainability measurement framework: The case of employee perspective in a CSR-committed organisation. *Journal of Cleaner Production, 188,* 708–719. https://doi.org/10.1016/j.jclepro.2018.03.269.

Sweeney, L. (2009). *A study of current practice of Corporate Social Responsibility (CSR) and an examination of the relationship between CSR and financial performance using Structural Equation Modelling (SEM)* (Doctoral thesis). Dublin Institute of Technology, Dublin.

Tucker, M. E., & Grim, J. (2007, Spring). Daring to dream: Religion and the future of the earth. *Reflections—The Journal of the Yale Divinity School, 4.*

Tyagi, R., & Raju, J. (2018, June). The effect of entrant brand's ownership on national brands' positioning strategies. *Management & Decision Economics, 39*(4), 475–485. https://doi.org/10.1002/mde.2919.

United Nations Global Compact, in collaboration with BCG. (2014). *Joining forces: Collaboration and leadership for sustainability* (Research Report January 2015). MIT Sloan Management Review.

United Nations—Global Corporate Sustainability Report (2013). Available from http://www.unglobalcompact.org/AboutTheGC/global_corporate_sustainability_report.html. Accessed on 24 February 2014.

Wackernagel, M. (2002). *Comments on draft GRI Sustainability guidelines.* http://globalreporting.org/feedback/PublicComments2002/MathisWackernagel1.pdf. Accessed on 14 February 2005.

Whelan, K. (2014, March). Ireland's economic crisis: The good, the bad, and the ugly. *Journal of Macroeconomics, 39*(Part B), 424–440. https://doi.org/10.1016/j.jmacro.2013.08.008.

Yan, Y., Wang, C., Quan, Y., Wu, G., & Zhao, J. (2018, March). Urban sustainable development efficiency towards the balance between nature and human well-being: Connotation, measurement, and assessment. *Journal of Cleaner Production, 178,* 67–75. https://doi.org/10.1016/j.jclepro.2018.01.013.

Zahra, S. A., Gedajlovic, E., Neubaum, D. O., & Shulman, J. M. (2009). A typology of social entrepreneurs: Motives, search processes and ethical challenges. *Journal of Business Venturing, 24,* 519–532.

Chapter 4
Financial Appraisal of a Micro-Generation Wind Turbine with a Case Study in Ireland

Abstract This chapter evaluates the economic benefits to be gained by installing a micro-generation wind turbine for a customer with a three-phase electrical supply requirement on an agricultural premise. The wind turbine is a 10-kW three-phase synchronous type and is embedded with the national electricity supply. It is anticipated that the energy units supplied by the wind turbine offsets, and reduces, the number of energy units imported from the National Grid. The evidence for the claims made in this paper is obtained by using actual empirical data collected from the installed equipment over three years. The objective is to accurately appraise the financial investment using real data downloaded from the renewable plant. There appear to be minimal studies conducted into this type of empirical research, possibly because the renewable energy sector is in the infancy stage in the host country, Ireland. There are some wind energy installations with financial appraisal techniques based on modelled data, which may, or may not, be accurate. The study concludes by claiming that the economic benefits of the wind energy turbine installation displayed disappointing results when compared to predicted benefits based on modelled data.

4.1 Introduction

The majority of Ireland's generated electricity comes from fossil-fuel-driven plants. In line with European Union directives, Ireland has committed itself to adjust this policy by agreeing on new climate and energy targets (Energy and Natural Resources 2018). There appears likely to be financial penalties to be paid by the Irish government if the targets are not met. It is hoped by the year 2020 that the renewable contribution to electricity production has increased to 40%. Of this figure, it is envisaged that 35% comes from wind energy. To aid and enhance this strategy, the Irish government has put incentives in place to encourage micro-generation wind energy projects. It appears that now a significant number of small businesses and households have embraced these types of wind energy projects, possibly without fully investigating the consequences of adopting such incentives.

Financial appraisals of micro-generation individual projects appear to be sparse, understandably because of the early stage of development of this industry life cycle.

© Springer Nature Switzerland AG 2020   83
T. Kealy, *Evaluating Sustainable Development and Corporate Social Responsibility Projects*, https://doi.org/10.1007/978-3-030-38673-3_4

A paper by Kelleher and Ringwood (2009) presents a method to estimate the economics of renewable micro-generation of electricity from wind and solar energy sources using a computer program. Kelleher and Ringwood (2009) use variables such as a range of feed-in tariffs, government incentive schemes, and the cost of capital borrowing to determine Payback Periods. They concluded by claiming that Payback Periods can vary greatly depending on the location, installation, and economic variables. Location is also seen as being a significant variable in an article by Al-Buhairi and Al-Haydari (2012). They highlight the importance of carrying out preliminary analysis on potential wind turbine sites as they claim that the amount of energy that can be supplied depends on the wind resource available, the type of wind turbines used, and the nature of the load being supplied. The methodology used (Al-Buhairi and Al-Haydari 2012) involved analysing wind speed data collected over seven years by the Yemen Meteorological Department.

As stated by Kelleher and Ringwood (2009), the economic variable is likely to influence the level and attractiveness of feed-in tariff available to the turbine owner(s). One such study was carried out by Walters and Walsh (2011) who examined the financial performance of micro-generation wind projects in the UK with a specific focus on the subsidy effect of feed-in tariffs. However, the benefits and cost savings of such projects in Ireland have yet to be identified using empirical data from existing installations. A case study carried out in Pakistan by Awan et al. (2012) questioned whether it is worth investing in infrastructure for wind energy alone and proposes that hydroelectric power be used in conjunction with wind power to reduce the variation in the output. Awan et al. (2012) mentioned the results of a survey of wind data in a range of potential sites. The proposed 600-kW wind turbine had a predicted energy output of 696,663 kWh units per year. While it is essential to consider the economic outcomes of a wind turbine investment, it is worth noting that the decision may not be made on a purely financial basis only. The public perception of such energy sources is also essential for energy policy. As stated by Burger and Gochfeld (2012), while renewable energy must be cost-effective, monitoring human perceptions of energy sources is also crucial for energy policy. Social attitudes change over time and are influenced by population density, technologies, and economic consequences.

This longitudinal ethnographic study on a 10-kW, three-phase, micro-generation, wind turbine took place on a singular farm unit in County Meath, Ireland, in 2012/2013. The annual electrical energy imported usage was 77,312-kWhs in 2009, 77,064-kWhs in 2010, 68,519-kWhs in 2011 and 76,338 kWhs in 2012. The wind turbine was commissioned in February 2010. It is a hypothesis of the study that the number of imported energy units would decrease after the wind turbine became operational. The researcher worked closely on-site, collecting first-hand experiences about the project.

4.2   Chapter Methodology

A case study methodology was embraced in this chapter. Initially, a site visit to the premises was arranged, enabling relevant quantitative data to be obtained from the electrical metering equipment. Subsequently, many electrical utility bills were accessed online in agreement with the turbine owner to examine the case in detail.

A. Evaluation criteria (Sect. 4.3)

The performance of the wind turbine installation was evaluated from the following perspectives:

(1) Initial cost (Sect. 4.3.1),
(2) Power output (Sect. 4.3.2),
(3) Energy output (Sect. 4.3.3) and
(4) Financial investment appraisal (Sect. 4.3.4).

In addition to the four evaluating perspectives, an electrical power output meter (Fluke 1735) was connected to the cables linking the inverters and the main distribution board. The resulting data graphically plotted over a 36-minute test period. From the resulting data, a measure of power dispersion was calculated.

B. Schematic diagram

The schematic diagram for the micro-generation wind turbine installation is shown in Fig. 4.1. It indicates the single-phase AC output from the left-hand inverter connected to L_1 while the AC output from the right-hand inverter connected to L_2, via an isolating transformer. The inverters are programmed so that the left-hand inverter has priority over the right-hand inverter and therefore produces an AC output at a lower DC input voltage level and produces a larger number of energy units. The schematic indicates that a 3-core, Steel Wire Armour (SWA) cable, buried directly in the ground linking the turbine generator and the farm installation, is to be sized

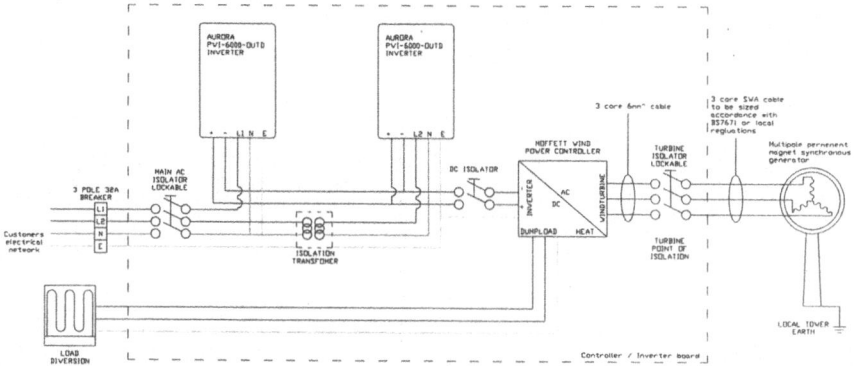

Fig. 4.1 Schematic diagram for 10-kW synchronous wind turbine installation

following British Standard BS7671 (2008) or local regulations. The distance between these two points is 300 m. The generator is a three-phase, multipole, synchronous micro-generator. Micro-generation refers to an electrical generator with a power output of less than 11-kW (DCCAE 2018).

4.3 Evaluation Criteria

4.3.1 10-kW Wind Turbine Initial Cost

The turbine installation cost was €22,000 plus Valued Added Tax at 21%, making the total price equal to €26,620. Maintenance of the installation is included in the initial cost. Due consideration must be given to the maintenance of the turbine, and it is suggested by Zaghar et al. (2012) that analysis of the failure rate of each component is fundamental at the design stage as maintenance is generally difficult and costly. The specification for the turbine is shown in Table 4.1.

This price included the supply and installation of a three-phase 12-kW inverter for the interface between the turbine and the existing electrical installation. However, during on-site inspection, it was found that the contracted installation company installed two single-phase six kW-rated inverters instead of the quoted three-phase version. The original quote also included the installation of a 25 mm² Steel Wire Armour cable, costing €6.45 per metre, to carry the current from the turbine to the facility. The installation company were new entrants in the renewable energy industry.

Table 4.1 Specification for synchronous wind turbine generator

Turbine type	Upwind
Rated capacity	10 kW
Maximum capacity	12 kW
Rotor diameter	6.5 m
Number of blades	3
Rotor speed	0–260 RPM
Generator type	Permanent magnet
Cut-in wind speed	2.2 m/s
Rated wind speed	11 m/s (39 km/h)
Cut-out wind speed	30 m/s
Survival wind speed	58 m/s (200 km/h)
Yaw control	Active
Main brake	Winch yaw control
Tower height	10 m
Performance	900–2100 kW per month

They made a strategic decision to enter the renewable energy market after successfully competing in a different sector for many years. Before installation began, there were no tests carried out to ascertain the suitability, or otherwise, of the site. Testing should have included wind speed tests at the proposed location of the turbine. Also, there were no load (current) checks carried out at the clients existing installation to determine if the electrical loads were balanced equally over the phases as recommended by Al-Buhairi and Al-Haydari (2012). The three-phase utility meter at the supply intake is equipped with both an Import and an Export facility. Any excess power generated from the turbine, and not used instantaneously on the farm, is exported onto the National Grid. The number of export units is 477 kWh units per annum. The farmer receives 9 cents per kWh for every unit of energy exported. The lifespan of the turbine in this research is quoted as being 25 years. Zaghar et al. (2012) suggest that the lifetime of turbines might be 20 years, but a final statement cannot be made because they are in the infancy stage.

4.3.2   Rated Power Output

The wind turbine has a rated capacity of 10-kW with a maximum output capacity of 12-kW. The generator is a multipole, permanent magnet, three-phase synchronous generator. The turbine has a rated wind speed of 11 m/s as specified in Table 4.1.

4.3.3   10-kW Wind Turbine Energy Output

Each inverter has an energy output indicator on the front panel. This data is recorded and used in subsequent calculations for this research. Over three years, the two single-phase inverters produced a combined total of 21,779 kWh units of energy (see 'Appendices for Chapter 4'). The left-hand inverter, Fig. 4.1, produced 13,307 kWhs and the right-hand inverter produced 8472 kWhs of this total. These values equate to average yearly energy output, for the turbine, of 7260 kWhs. Of this annual total, 477 kWh units of energy are exported back to the National Grid at a feed-in tariff rate of 9 cents per kWh. This rate gives a net import energy saving of 6783 kWhs per annum. As a result of examining previous utility bills over many years, it is noted that the customer uses 55% of his electricity during the day and 45% at night. Therefore, the actual imported energy savings are 55% of 6783 (3731 kWhs) day units and 45% of 6783 (3052 kWhs) night units. Note that the performance specification shown in Table 4.1 predicts an energy output of 900–2100 kWh per month (10,800–25,200 kWh per annum). A summary of the yearly savings is as shown in Table 4.2.

Table 4.2 Savings made due to 10-kW wind turbine installation

Day units	Day rate (€)	Night units	Night rate (€)
3731 kWh	0.1815	3052 kWh	0.0897
	677		274
Plus VAT	91	Plus VAT	37
Subtotal	788	Subtotal	311
Export 477 kWh at 9 cent per kWh = €43			
Total annual financial benefits = €1142			

4.3.4 10-kW Turbine Financial Investment Appraisal

The turbine installation was a significant investment by the farmer. Given the importance of this investment decision, it is essential to screen the investment proposal. There are four main methods of evaluation used in this research (Atrill and McLaney 2013). They are (1) Payback Period (PP); (2) Accounting Rate of Return (ARR); (3) Net Present Value (NPV); and (4) Internal Rate of Return (IRR).

Payback Period (1): This is the length of time it takes for the initial investment of €26,620 to be repaid out of the net cash inflows from the turbine installation. We can derive the Payback Period by calculating the cumulative cash flows associated with the project. The cumulative cash flow becomes positive after year twenty-three, as shown in Table 4.3.

The advantages of the PP method are that it is quick and easy to calculate and is easily understood by the manager making the investment decision.

Accounting Rate of Return (2): This investment appraisal method takes the average accounting operating profit that the wind turbine installation generates and

Table 4.3 Payback Period for 10-kW wind turbine

Time	Net cash flow (€)	Cumulative cash flow (€)
Immediately	−26,620	
1 years' time	1142	−25,478
2 years' time	1142	−24,336
3 years' time	1142	−23,194
4 years' time	1142	−22,052
5 years' time	1142	−20,910
21 years' time	1142	−2638
22 years' time	1142	−1496
23 years' time	1142	−354
24 years' time	1142	788
25 years' time	1142	1930
25 years' time	2000	3930

expresses it as a percentage of the average investment made over the lifetime of the project, i.e. 25 years. The average annual operating profit is the cash flow (€1142) plus the depreciation on the installation (€26,620/25, i.e. €1064.80) giving a total value of €2206.80. The average investment is the cost of the plant plus the scrap value, all divided by two ((€26,620 + €2000)/2) giving a value of €14,310. The ARR of the turbine installation is calculated as 15.42% [(€2206.80/€14,310) × 100%]. The ARR relates accounting profit to the cost of the assets invested to generate that profit. The problem with ARR is that it almost completely ignores the time factor. There are also problems concerning the approach taken to derive the average investment of the turbine.

Net Present Value (3): The NPV investment appraisal method considers all of the costs and benefits of the micro-generation wind turbine installation and makes a logical allowance for the timing of these costs and benefits. The time factor is an essential factor as the farmer does not see €1142 received now as equivalent in value to €1142 receivable in a year. The three reasons for this are (1) interest lost; (2) risk; and (3) effects of inflation. The NPV method makes a direct comparison between the sum of the inflows over time and the immediate €26,620 investment. The cash benefits over time are discounted, depending on the interest rate and the period (year) in which the benefits arise. The NPV value for an initial discount factor of 13% is shown in Table 4.4.

Based on a discount factor of 13%, the NPV of the wind turbine project is − €18,215. The decision rule for NPV states that if the NPV is positive, the project

Table 4.4 Net present value for 10-kW wind turbine

Time	Cash flow (€)	Discount factor (13%)	Present value (€)
Immediately	−26,620	1	−26,620
1 years' time	1142	0.885	1011
2 years' time	1142	0.783	894
3 years' time	1142	0.693	791
4 years' time	1142	0.613	700
5 years' time	1142	0.543	620
6–21 years'			
22 years' time	1142	0.065	74
23 years' time	1142	0.060	69
24 years' time	1142	0.053	61
25 years' time	1142	0.047	54
	2000	0.047	94
		NPV	**−18,215**

should be accepted, and if the NPV is negative, the project should be rejected. The NPV method seems to be a better method of appraising the wind turbine installation because it takes into account the following three criteria: (1) the timing of the cash flows; (2) the whole of the relevant cash flows; and (3) the objectives of the business (Atrill and McLaney 2013). In this case, it would appear that investment in the project is not viable because the NPV is a negative value, indicating that the costs outweigh the benefits. This NPV value in Table 4.4 is based on a discount factor of 13%. The wind turbine project was funded by the business owner through a loan only arrangement. The bank loan interest rate was 4%. The new NPV for the project now calculates at a value of −€9380, as shown in Table 4.5, for a 4% discount factor.

The NPV calculations shown in Tables 4.4 and 4.5 demonstrate the sensitivity analysis of the interest rate input variable for the financing of the project. The complete list of all values for the 4% interest rate for Table 4.5 is shown in Table A4.2 'Appendices for Chapter 4'. Another sensitivity analysis calculation is carried out on the project by assuming an interest rate from a financial institution of 7% as the rate is based on the cost of borrowing capital. When the interest rate is 7%, the NPV for the project is −€13,122 still indicating that the costs outweigh the benefits.

Sensitivity analysis on 10-kW Wind Turbine Capacity Factor—The actual Capacity Factor (CF) for the turbine in this case study is approximately 8% (7260/87,600). This CF value is quite low in comparison to other wind turbine installations (Henaghan 2013). At this 8% CF value, 7260 kWh units of electrical energy are produced each year and yield a saving of €1142 annually (Table 4.2). Based on the

Table 4.5 NPV for 4% interest rate for 10-kW wind turbine	Time	Cash flow (€)	Discount factor (4%)	Present value (€)
	Immediately	−26,620	1	−26,620
	1 years' time	1142	0.96	1096
	2 years' time	1142	0.925	1056
	3 years' time	1142	0.89	1016
	4 years' time	1142	0.855	976
	5 years' time	1142	0.82	936
	6–21 years'	In 'Appendices for Chapter 4'		
	22 years' time	1142	0.42	479
	23 years' time	1142	0.405	462
	24 years' time	1142	0.39	445
	25 years' time	1142	0.375	428
		2000	0.375	750
			NPV	**−9380**

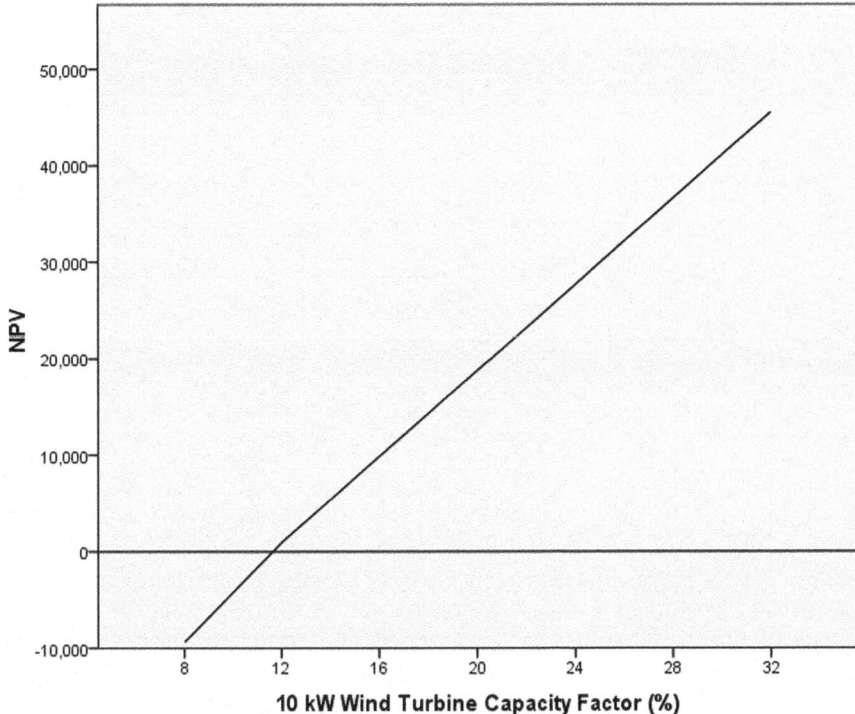

Fig. 4.2 Sensitivity analysis for 10-kW turbine using a range of capacity factor values between 8 and 32% (*X*-axis)

actual 4% interest rate for the capital loan, the NPV is calculated using improved CF values of 12, 16, 20, 24, 28, and 32%. The results are shown in Fig. 4.2.

Internal Rate of Return (4): The IRR method of investment appraisal, like NPV, involves discounting future cash flows. The IRR of the wind turbine installation is the discount rate that, when applied to its future cash flows, produces an NPV of precisely zero. In essence, it represents the yield from the turbine investment. From (3), we calculated the NPV of the installation at an interest rate of 13% as −€18,215. When the interest rate is set at 2%, the NPV is calculated as −€3110. When the interest rate is set at 1%, the NPV is calculated at €80. Since the IRR is the discount rate that gives an NPV of precisely zero, we can conclude that the IRR of the installation is between 2 and 1%. A more accurate calculation is 1.025%. The IRR calculation is shown in Table 4.6.

It is important to note that the methods described, and the values calculated are not seen purely as a mechanical exercise. The results derived from this wind turbine installation investment appraisal are only one input to the decision-making process. Other, broader, issues that may be connected to the decision include the concern, by the farmer (investor), for our natural environment which, according to much scientific evidence, appear to be under the threat of global warming. Each kWh unit of

Table 4.6 Internal rate of return for 10-kW wind turbine

Discount factor (2%)	Present value (€)	Discount factor (1%)	Present value (€)
1	−26,620	1	−26,600
0.98	1119	0.99	1131
0.961	1097	0.98	1119
0.942	1076	0.971	1109
0.924	1055	0.961	1097
0.906	1035	0.951	1086
0.647	739	0.8	914
0.634	724	0.795	908
0.622	710	0.787	899
0.609	695	0.779	889
0.609	1218	0.779	1558
NPV	−€3110	NPV	€80

energy produced by the turbine offset 0.530 kg of CO_2 that would be emitted into the atmosphere as a result of burning fossil fuel to produce the same output (2010 value, Table A4.1). The carbon intensity value of 0.530 kg CO_2 is the simplified average amount of CO_2 offset based on the SEAI benchmarks for that year (Table A4.1). Carbon intensity is the amount of carbon dioxide that is released to produce one unit (kWh) of energy for a given fuel (Table A4.1). The carbon intensity to create one unit of electricity depends on the fuel mix used to generate electricity and also on the efficiency of the technology employed (Sect. 1.1.2). The average plant emissions rates may not reflect precisely the emission savings from wind generation as the exact value depends upon the type of generator used to increase/decrease its output in response to the varying wind turbine speed. This wind turbine output variation is frequently met by gas generators which are designed to operate at more variable levels of output and can be cycled quickly (Kaffine et al. 2013). The variations in the power output from the 10-kW three-phase synchronous wind turbine are demonstrated graphically in Fig. 4.3. The parallel-connected traditional, embedded generators must be capable of accommodating the rapidly varying power output from the 10-kW synchronous generator. The turbine power output variations are expressed as Coefficient Of Variation (COV) values. Large COV values increase the plant emission rates of backup, embedded traditional generators.

The COV values for the data presented in Fig. 4.3 are 0.6 (60%, expressed as a percentage). The test period is thirty-six minutes long, and the sampling time is 0.5-s. The high dispersant power is observable from Fig. 4.3.

It is hoped that we, at this present time, do not destroy the natural environment to be inhabited by future generations because of our heavy dependence on burning imported fossil fuels. This ecological evaluation of renewable energy sources is summarised by Burger and Gochfeld (2012) who list seven objectives that must be met to make renewable energy effective. The authors claim that renewable energy

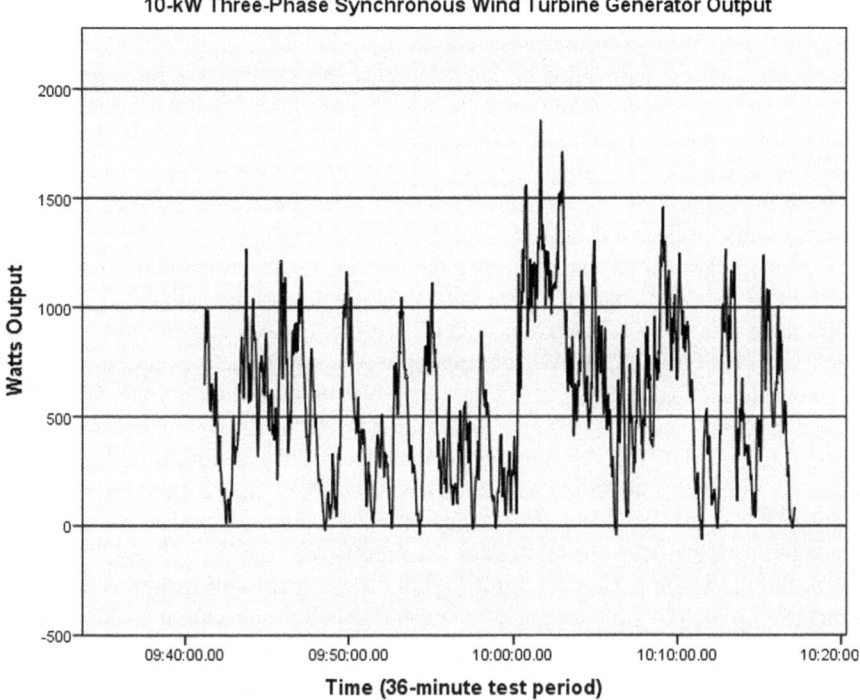

10-kW Three-Phase Synchronous Wind Turbine Generator Output

Fig. 4.3 Short-term variations in 10-kW generator power output on 14 April 2016

Table 4.7 Summary of financial appraisal methods for 10-kW synchronous wind turbine

Appraisal method	PP	ARR	NPV	IRR
Value	23 years	15.42%	−€18,215	1.025%

must be protective of human health and the environment, protective of landscape and Earth systems, and be acceptable to the public. A summary of the results of the financial appraisal methods for the wind turbine installation is expressed in Table 4.7.

4.4 Findings

A potentially significant outcome of the case study was highlighted by measuring the load current at the Distribution System Operator (DSO) electrical supply intake of the installation. It was found that a possible inefficiency in the design of the facility may have negatively affected the potential for savings on the project. It was noted that the output from the left-hand inverter was connected to L_1 of the plant and from the right-hand inverter to L_2, as shown in Fig. 4.1. The only connection to L_3 was via the National Grid. However, on analysis of the loads connected to the installation, it was discovered that L_1 was the phase with the lightest loads connected to the supply. The problem was compounded because the left-hand inverter was programmed to give the highest output of the two inverters. The result was that the farmer could potentially be exporting electricity via L_1 at 9 cents per kWh and, at the same time, importing electricity on either L_2 or L_3 at 18 cents per kWh. As a result of this analysis, the output from the left-hand inverter was moved to L_2, and the output from the right-hand inverter was moved to L_3.

The researcher found that the cable buried directly in the ground, installation method D (British Standard BS 7671 2008) linking the turbine with the installation is 3-core 25 mm^2 SWA. When the cable is carrying, for example, 30% of the rated output from the turbine, 3.3-kW, this equates to a current value of approximately 13 A. Under these conditions, the total volt drop between the start and the end of the cable is 5.85 Volts (1.5 mV × 13 × 300). When the full load is being generated, i.e. approximately 40 A, the volt drop in the cable is 18 Volts, a significant loss in the cable. A cable with a larger diameter would have reduced these losses considerably. For example, a similar cable with a conductor size of 35 mm^2 would have a volt drop on the full load of 13.2 Volts (1.1 mV × 40 × 300).

The turbine was found to have significant short-term dispersion in the power output signal (Fig. 4.3). The wind turbine was embedded in the agricultural installation on two of the three electrical supply phases (Fig. 4.1). The (utility) electricity supply must compensate for the fluctuating turbine output. The electric utility is supplied by a mix of generating technologies, but much of the supply is fossil-fuel-driven generators. Ramping of fossil-fuel generators affects the emission rates, and therefore efficiency, of the generating plant (Cullen 2013).

4.5   Discussions

From this ethnographic/case study, it appears that the expected economic benefits of investing in this micro-generation wind energy project did not materialise. The client made a significant financial investment, €26,620, to reduce the overall electrical energy bill by only 9.5% (7260/76,338). The results of the values calculated by the financial appraisal methods are disappointing. A PP of 23 years is unlikely to be acceptable to shareholders in the business. As a comparison to generally acceptable economic benefits from investment opportunities, some examples are now briefly discussed. The supermarket giant, Tesco, is in the process of installing voltage optimiser equipment at the DSO intake to nearly all of its 2300 stores and warehouses in the UK (Jaggi 2009). The optimiser reduces the voltage, if required, to allow connected loads run at optimum efficiency. Tesco expects a return on investment of approximately 20% and achieves a Payback Period of five years by installing the voltage optimiser equipment. Also, Marks & Spencer, the stores' chain, has targeted an IRR of between 12 and 15% on any new investment programme (Marks & Spencer 2011). Thirdly, Rolls-Royce, in its 2010 annual report and accounts stated that all investments are subject to a rigorous examination of risks and future cash flows to ensure that they create shareholder value (Rolls-Royce 2010). Discounted cash flow (NPV) analysis is performed regularly at Rolls-Royce. The Payback Period of the turbine in this research is significantly more extended than that predicted in the study by Kelleher and Ringwood (2009). For example, Kelleher and Ringwood predict a 3.65 years Payback Period for a proven 2.5-kW micro-generation wind turbine in an open rural area. It must be noted, however, that the range of sizes used in Kelleher and Ringwood's research is smaller than the turbine used in this research. There appear to be several factors contributing to the indication that financially, the wind turbine project does not perform well in this case.

Firstly, the competencies of some companies' operating in this specialised area would seem to be somewhat questionable. It appears that the installation company in this research did not have the expertise needed to design and install such an installation. They did not complete any pre-connection wind speed or electrical load tests on the plant, as suggested by Al-Buhairi and Al-Haydari (2012) and they did not inform the client of the potential pitfalls, or advantages, that his investment might hold. This conclusion concurs with Walters and Walsh (2011) who claimed that how the equipment is installed contributes to the success, or otherwise, of the project. In the installation of the wind turbine for this research, two single-phase inverters were installed instead of a three-phase inverter, which may contribute to a lower energy output than specified by the manufacturer. There was significant dispersion in the turbine output signal shown in Fig. 4.3. The COV was calculated at 0.6 (60%) for the data demonstrated in Fig. 4.3. This large dispersant value affects the parallel-connected fossil-fuel generators, which must attempt to compensate rapidly to supply stable power to the connected loads, as stated in Sect. 4.3.4. In addition to the dispersion issue, the SWA underground cable linking the turbine generator and the installation appeared to be lower than that needed to efficiently transfer the power between both, considering the distance is significantly longer at 300 m. This long distance leads to high power loss in the cables.

Secondly, the renewable energy feed-in tariff, at 9 cents per kWh, is low compared to UK tariffs. Walters and Walsh (2011) concluded that the proposed feed-in tariff of 30.5 (UK) pence per kWh (approximately 25.9 cents per kWh) would not boost the economic attractiveness of some sites in the UK. There seems little benefit, in Ireland, of customers exporting electricity at a significantly lower price per unit than the UK when the higher price is deemed unattractive in the UK. The customer in this study is better advised, from an economic point of view, to use all of his generated units in his installation rather than export any to the National Grid. Table 4.8 shows the benefits to the consumer if all the electricity generated by the turbine is used on-site in the installation. We can compare these results with the figures in Table 4.2. When the financial benefits are analysed, allowing for 477 kWh units to be used on-site rather than be exported to the National Grid as shown in Table 4.2, it can be concluded that there is a minimal financial benefit to be gained by using all the generated kWh units on the installation, as shown in Table 4.8. The difference in monetary terms is €14 per annum.

Table 4.8 All 10-kW wind turbine generated kWh units used on-site

Day units	Day rate (€)	Night units	Night rate (€)
3993 kWh	0.1815	3267 kWh	0.0897
	725		293
Plus VAT	98	Plus VAT	40
Subtotal	823	Subtotal	333
Total annual financial benefits = €1156			

Thirdly, it may be significant that the specialised, and new, nature of these wind energy projects are such that in many cases a client's understanding and knowledge of the venture, its terminology and the technology involved are somewhat limited and therefore the potential for exploitation is great. The investor in this research used his 'gut feeling' in making this investment decision. Larger businesses can afford to employ financial experts to appraise any such potential projects.

However, on the positive side, this investment offset a total of 3847.8 kg of CO_2 being emitted into the atmosphere (7260×0.530) annually as a result of 'green' generation of electrical energy instead of burning fossil fuels to obtain the same output (2010 value, Table A4.1). This can contribute to Irish efforts to reach energy targets as set by the European Union Directives concerning CO_2 emissions and possibly reduce the fines to be paid by the Irish government if the goals are not met.

4.6 Conclusions

The financial analysis of the micro-generation wind turbine investment decision identified disappointing results in this case study. A long Payback Period and a negative NPV value would seem to indicate that the cost of the investment outweighs the financial benefits to be gained by making the investment. The disappointing financial results attributed to the sustainability initiative in this case study are linked to the short-term, highly dispersant, turbine output signal. The embedded, backup, generators, many of which are fossil-fuel-driven, are unable to respond quickly enough to compensate for the fluctuations. The result is that there is minimal reduction in the year-on-year energy imported values for this investor as specified for 2009/2010/2011/2012 in Sect. 4.1. As the turbine was commissioned in February 2010, a significant reduction in imported energy units was expected after that date.

There were no preliminary tests carried out on the suitability of the site or the load characteristics of the installation. However, the Irish government is keen that such renewable energy projects are implemented as it contributes to reducing CO_2 emissions. This helps the country to meet predefined targets; otherwise, financial penalties may be incurred.

The main contribution of this research is to provide an appraisal of a micro-generation wind turbine installation using on-site measured data collected over several years and the results of which can be used for future, potential, investors in their investment decisions. There is a lack of such research in the host country, Ireland, and uncertainty surrounds the methods by which similar projects are economically, and environmentally, evaluated. These results assist in the decision-making of other potential investors.

The author proposes that there is merit in carrying out an investigation on a larger wind project, to include measurement of the renewable generator power output dispersion value, where the designer/installer has expert knowledge and experience in the wind energy industry.

References

Al-Buhairi, M. H., & Al-Haydari, A. (2012). Monthly and seasonal investigation of wind characteristics and assessment of wind energy potential in Al-Mokha, Yemen. *Journal of Energy and Power Engineering, 4,* 125–131.

Atrill, P., & McLaney, E. (2013). Accounting and finance for non-specialists (8th ed.). Harlow: Pearson Education.

Awan, S., Ali, M., Asif, M., & Ullah, A. (2012). Hydro and wind power integration: A case study of Dargai station in Pakistan. *Journal of Energy and Power Engineering, 4,* 203–209.

British Standard BS7671. (2008). *Requirements for electrical installations* (7th ed.). IEE Wiring Regulations, The Institution of Engineering and Technology and BSI.

Burger, J., & Gochfeld, M. (2012). A conceptual framework evaluating ecological footprints and monitoring renewable energy: Wind, solar, hydro, and geothermal. *Journal of Energy and Power Engineering, 4,* 303–314.

Cullen, J. (2013). Measuring the environmental benefits of wind-generated electricity. *American Economic Journal: Economic Policy, 5*(4), 107–133. https://doi.org/10.1257/pol.5.4.107.

DCCAE. (2018). Department of Communications, Climate Action & Environment. Available at http://www.dccae.gov.ie. Accessed on 23 May 2018.

Energy and Natural Resources. (2018). http://www.dcenr.gov.ie.

Henaghan, D. (2013). *Analysis of the capacity factor of Irish wind energy*. Dissertation as part of the MSc in Energy Management (DT711). Dublin Institute of Technology.

Jaggi, J. (2009, November 25). Case study: Power efficiency. *The Financial Times.* Retrieved from http://www.ft.com/intl/cms/s/0/c882555a-d9e6-11de-b2d5-00144feabdc0.html#axzz312a14MVA.

Kaffine, D. T., McBee, B. J., & Lieskovsky, J. (2013). Emissions savings from wind power generation in Texas. *The Energy Journal, 34*(1). https://doi.org/10.5547/01956574.34.1.7.

Kelleher, J., & Ringwood, J. V. (2009). A computational tool for evaluating the economics of solar and wind micro-generation of electricity. *Energy, 34*(4), 401–409.

Marks & Spencer plc. (2011). *Annual report*. London: Marks and Spencer.

Rolls-Royce plc. (2010). *Annual report*. London: Rolls-Royce.

Walters, R., & Walsh, P. R. (2011). Examining the financial performance of micro-generation wind projects and the subsidy effect of feed-in tariffs for urban locations in the United Kingdom. *Energy Policy, 39*(9), 5167–5181.

Zaghar, H., Sallaou, M., & Chaaba, A. (2012). Preliminary design support by integrating a reliability analysis for wind turbine. *Journal of Energy and Power Engineering, 4,* 233–240.

Chapter 5
Post-connection Financial Performance Analysis of a Four-Turbine, 3.5-MW, Wind Farm in Ireland

Abstract With many electricity markets worldwide deregulated or in the process of deregulation, the opportunity for smaller independent generators to provide power to their local power system has increased. For smaller independent wind developers assessing the feasibility of a large-scale wind farm project is vitally important due to significant risk associated with the investment. This paper presents a longitudinal case study of a 3.5-MW wind farm situated in the North East of Ireland, utilising multiple sources of empirical data obtained over three years following commissioning. The findings indicate that an average yearly capacity factor of 34% was recorded from the turbines providing for a simple payback period of 6.7 years. It would appear from this case study that site selection, electricity market conditions, the quality of the control system and the competencies of the design/installation/commissioning company all contributed to the satisfactory results.

5.1 Introduction

The relentless push for business growth has put pressure on natural and human resources. For example, the supply of oil, coal, and gas would appear to be at or approaching, the end period of their life cycle. In addition to this, many scientists argue that the burning of fossil fuels in power generating plants is contributing to environmental degradation, which over time, could jeopardise our wealth, and even our existence (Catalin and Nicoleta 2011). This fragile relationship between the economy and the environment has been primarily ignored as business growth, and shareholder profits took precedence over the natural environment. In more recent times research papers, reports, and international conferences such as Kyoto (1997) and Doha (2012) have drawn attention to an imminent economic, environmental, and human crisis. From these studies, a new model of business management has emerged, namely sustainable development, which links the economic, environmental, and social spheres. The wind energy industry grew out of this shift in business thinking, as wind turbines were considered as an alternative to fossil fuel burning plants for power generation. However, caution needs to be applied as there appears to be minimal empirical data with which to compare *actual* Payback Periods with

© Springer Nature Switzerland AG 2020
T. Kealy, *Evaluating Sustainable Development and Corporate Social Responsibility Projects*, https://doi.org/10.1007/978-3-030-38673-3_5

predicted Payback Periods. One such study by Kealy (2014) identified a Payback Period of twenty-three years, a very disappointing result for potential wind energy investors. There are interplay and synergy between the economic, environmental, and social dimensions (Table 3.1). This interplay means that projects that benefit the investors financially *and* help to protect the environment contribute to the social (human) aspect of sustainable development also.

5.2 Literature Review

Ireland is committed to reducing its dependence on fossil fuels in line with European Union directives (Department of Communications, Energy and Natural Resources, DCENR 2015). The Irish government has initiated a policy whereby 40% of its electricity is intended to be generated by renewable sources by the year 2020 (SEAI 2013). Of this figure, it is predicted that a significant portion comes from wind energy which has prompted increased interest in wind farm development. If this wind energy plan materialises, Ireland can become one of the world's largest power-from-wind producers as a percentage of the total supply. While this study analyses the financial aspects of wind farm operation (utilising digital energy measuring instruments to measure the quantity of electrical energy production, Fig. 8.3), it is essential to remember that there are other environmental aspects associated with wind turbine design, which are not part of the remit for this current chapter. Some of these different aspects are discussed in a cradle-to-grave study of a four-turbine wind farm in Greece by Abeliotis and Pactiti (2014). While Abeliotis and Pactiti (2014) conclude that wind power is environmentally preferable compared to the current Greece generation mix, mainly fossil-fuel-driven plant, wind power is not entirely environmentally impact-free, since they consume raw materials and energy for their manufacturing, transportation, installation, maintenance, dismantling, and disposal. The holistic methodology applied by Abeliotis and Pactiti (2014) is called Life Cycle Assessment (LCA).

As part of Ireland's deregulated Single Electricity Market (SEM), private wind developers can construct and provide power to their local power system. However, caution needs to be applied when making predictions concerning potential energy output from these machines, which are in their infancy stage in the overall life cycle of the development of the wind turbines. There appears to be a shortage of publicly available empirical research to evaluate turbine performance based on data obtained from actual wind turbines installations, although there is a plethora of research modelling estimated values. One such estimated wind energy potential was investigated using locally accessible wind data for a potential site in Konya, Turkey. The one-year wind data was statistically analysed using computer software. A predicted simple Payback Period of 6.44 years for a 6-MW wind farm was calculated by Kose et al. (2014). It is anticipated that this Payback Period would attract local investors to invest in wind energy technology. In terms of actual data, an empirical study by Kealy (2014) investigated the financial performance of a 10-kW rated small-scale wind turbine installation in Ireland over the period from 2010 to 2013. The project had a capital cost of €26,620 and the turbine energy output was 7260 kWh units per annum. This energy output value yielded a Capacity Factor (CF) of just 8.3%. Considering that the client had an annual energy usage of approximately 76,338 kWh's, the investment did not put a significant reduction in his electricity usage, and the results appear disappointing. The Payback Period for the 10-kW turbine investment was approximately twenty-three years. Research by Henaghan (2013) analysing the Capacity Factor of Irish wind energy found that some companies were overestimating the energy output from these projects. The findings were based on data from 77 wind farms in both the Republic of Ireland and Northern Ireland in the period 2008–2012. Of the wind farms surveyed, the highest recorded CF was 38.48%, the lowest CF was 19.36%, and the average CF was 27.9%. This data was extrapolated from a Single Electricity Market Operator (SEMO) source. International Standard IEC 61400-12-1 (*Wind Turbines—Part 12-1: Power performance measurements of electricity producing wind turbines*) does provide a uniform methodology which should ensure that the measurement and analysis of turbine power performance are consistent, accurate, and reproducible. The power performance characteristics of the wind turbine are determined by the measured power curve and the estimated annual energy production.

The success or failure of wind turbine investment decisions ultimately depends upon many factors. Therefore, for smaller independent wind developers assessing the feasibility of a large-scale wind farm project is a significant undertaking due to the inherent risk associated with the investment. This chapter presents a longitudinal case study of a 3.5-MW wind farm situated in the North East of Ireland utilising empirical data obtained over three years (2010/2011/2012) following commissioning. The findings should be of benefit to wind developers assessing the feasibility of similar projects.

5.3 Methodology

5.3.1 Case Study

This longitudinal case study research project utilised multiple sources of data requested and obtained from the owners of the company involved in the production and sale of wind-generated electricity for the three years, 2010/2011/2012. The data included wind speed, energy output, loan repayments, and turbine availability of each turbine in the wind farm installation. The case study methodology was used as an in-depth in-context analysis of the financial benefits were explored to assist potential future investors in the wind energy industry. Three years of data were utilised in the assessment of the project. The researcher worked closely with the wind farm owners throughout the duration of the study. Several site visits took place during which observations and measurements were recorded. Site meetings were organised with the project owners where financial reports and power/energy output readings were analysed. Informal discussions also took place during these site meetings regarding aspects of the wind farm investment. The cost of the project and the financing structure were discussed with the owners and results are shown in Tables 5.3 and 5.4, respectively. Based on average empirical data recorded over the three years, a unit price of 8.2 cents per kWh unit of energy was used, and a Capacity Factor of 34% was calculated, and the results are reported in the 'Results' section. The results of this case study are based on electronic digital energy meters installed on-site in the electrical substation (Fig. 5.2). The 10-kV electrical substation, recently upgraded to 20-kV, is shown to the right-hand side of the diagram in Fig. 5.2. The turbine was available for 97% of the time, and the turbine efficiency was also 97%. Although it is not possible to generalise all wind farms based on one case study, these findings should help to augment the gap in knowledge of wind farm empirical research.

5.3.2 3.5-MW Wind Farm Site and Plant

The wind farm on which this chapter is based on is situated in the North East of Ireland. Before the investment decision being made, preliminary tests were carried out to determine the suitability of the site, some of which are now briefly described. It was advantageous that there was no forestry or dwellings located on-site. Wind data analysis was carried out by installing a portable anemometer. An environmental impact assessment took place, which included analysing the effects of the wind turbines on fauna and flora. Subsequently, a decision was made to progress with the investment. The wind farm was constructed in two phases with an estimated life span of twenty years. Phase one consisted of two Vestas V52 850-kW wind turbines. Phase 2 consisted of the installation of two Enercon E44 wind turbines, each with an output of 900-kW. Phase 1 had a pre-connection predicted capacity factor of 40.29% and Phase 2 had an expected capacity factor of 36.24%. The total power output of the wind farm is 3.5-MW. A picture of the site is shown in Fig. 5.1 and the on-site electrical layout is shown in Fig. 5.2.

Fig. 5.1 Picture of 4-turbine 3.5-MW wind farm site (Kealy)

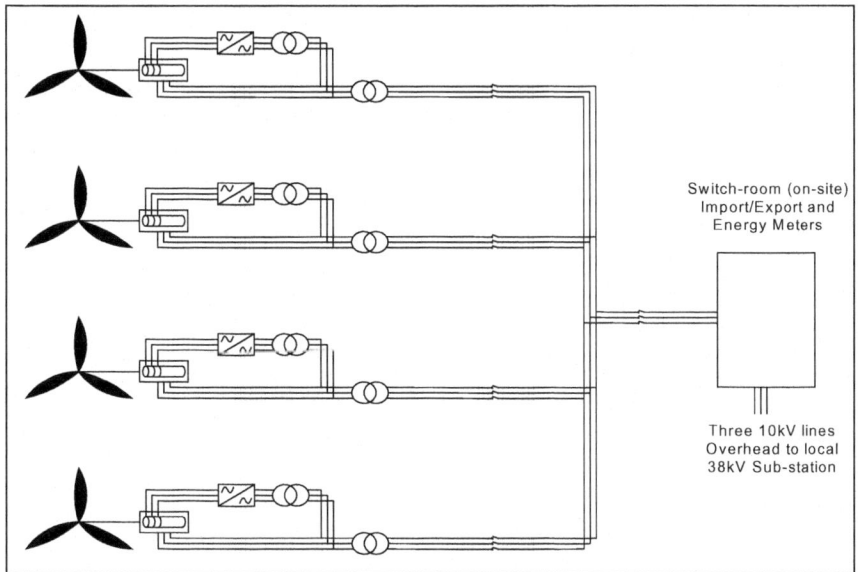

Fig. 5.2 Layout and electrical connections of four turbines in 3.5-MW wind farm

The most commonly used generator is the induction generator, of which the types include cage, wound rotor, and Doubly-Fed-Induction-Generator (DFIG). Each of the turbines assessed in this research paper is of the DFIG type. At the bottom of each turbine tower, a 690 V/10 kV transformer is installed (tapping on transformer recently changed to 690 V/20 kV to accommodate the national electrical grid upgrading), the output of which are coupled together and fed in underground cables to an on-site switch-room (Fig. 5.2). The switch-room is located 4 km from the 38-kV substation. The 20-kV wires are fed into the 'Wind Farm' node to the bottom right-hand-side of the drawing in Fig. 5.3. The 'Wind Farm' generated electrical energy units are used locally to Load A, Critical Load B, Load C, and Load D (Fig. 5.3). There are two 38-kV input supplies (Supply A and Supply B) and Critical Load B can be fed off either supply. It is expected that there is a significant reduction in the number of electrical energy units supplied through 38-kV Supply A and 38-kV Supply B because of the wind turbine parallel-connection (Fig. 5.3).

5.3.3 38-kV Local Substation

A significant upgrade of the 38-kV Electricity Supply Board substation was under-taken in 2012 to accommodate the 3.5 MW wind farm project (and also a new 3-MW turbine on an adjacent site in the planning stage). This upgrade included replacing an existing 2-MVA transformer with a 10-MVA transformer (T421). A single line

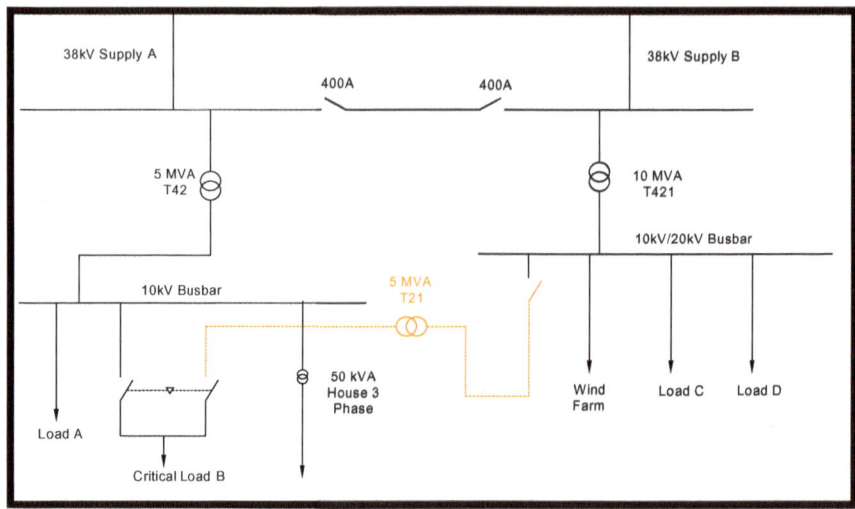

Fig. 5.3 Single line diagram for 38-kV substation

diagram of the 38-kV substation is provided in Fig. 5.3 with 'Wind Farm' connection to the bottom right of Fig. 5.3.

5.4   Overview of the Associated Plant for 3.5-MW Wind Farm

The control strategy to maximise the wind energy captured in a variable speed wind turbine with an internal induction generator at low to medium speeds is an essential aspect to the outcomes in wind energy projects, Iyasere et al. (2012). In a research paper, the authors, Iyasere et al. (2012) propose that the tip-speed ratio is controlled via the rotor angular speed, to an optimum point at which the power coefficient is at a maximum for a particular blade pitch angle and wind speed. The control systems used for the variable speed wind turbines in this research are also modern control systems, described in this section. The internal connections for both types of wind turbines in this project are shown in Fig. 5.4. Figure 5.4 shows the original 690 V/10-kV transformer feeding the local electricity 38-kV substation. This on-site transformer has subsequently been changed to 20-kV output in line with the Irish electricity utility infrastructure upgrade (Table 1.1).

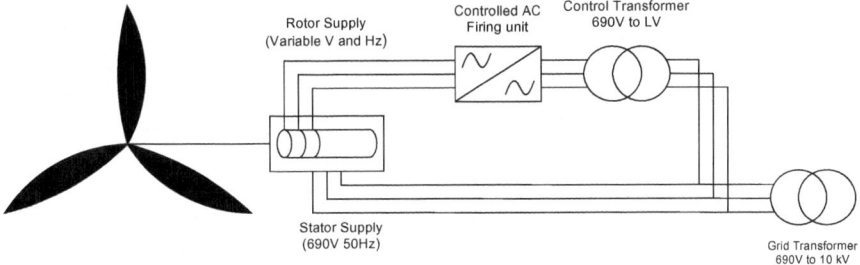

Fig. 5.4 Control system for each DFIG wind turbine

5.4.1   Design of the VESTAS V52–850-KW Wind Turbine

The Vestas V52–850-kW has a rotor diameter of 52 m and operates using the OptiSpeed™ concept. This feature enables the rotor to manage with variable speed (Revolutions Per Minute). These are also equipped with OptiTip®, the unique pitch regulating system. This system allows the angle of the blades to be continuously regulated so that they are always pitched at the optimal angle for current wind conditions. This control technique optimises power production and noise levels. The main shaft transmits the power from the rotating turbine blades to the generator through the gearbox, as shown in Fig. 5.4. The generator is a particular asynchronous four-pole generator with a wound rotor. OptiSpeed™ is also named Vestas Converter System (VCS), and this system ensures a steady and stable electric power from the turbine. The VCS consists of the following:

- A capable asynchronous generator with wound rotor and slip rings (Rotor supply in Fig. 5.4)
- A power converter with Insulated Gate Bipolar Transistor (IGBT) switches (Firing unit in Fig. 5.4)
- Contactors and protection (Overload and Short-circuit protection, not shown in Fig. 5.4)

The VCS enables variable speed operation in a range of approximately 60% of nominal Revolutions Per minute (RPM). It controls the current in the rotor circuit in the generator via slip rings. This technique gives full control of the reactive power and provides an accurate and precise connection between the generator and the National Grid. The generator stator is coupled directly to the 690-V supply. The wind turbine controller continuously collects data to control the performance of the turbine. Data continuously collected includes the following:

- Rotor and Generator speed,
- Wind speed,
- Hydraulic pressure,
- Temperatures,
- Power and Energy production,
- Pitch.

Table 5.1 Vestas nameplate on 850-kW ground controller in tower

Vestas wind systems	
Wind turbine type	V52–850-kW
Controller type	VMP–850-kW–90 V–50 Hz
Voltage	3×690 V+10/−10%
Frequency	50 Hz +1/−3 Hz
Current Cosθ = 1	711 A
Current Cosθ = 0.95	749 A
Max short-circuit current	Ik = 15 kA

The nameplate on the Vestas ground controller is shown in Table 5.1.

The generator is coupled in star mode if the total power output is low and delta mode if there is high power output from the generator. The VCS frequency converter is a four-quadrant converter which can provide a current in any direction and frequency on the grid side (grid inverter) and the rotor side (rotor inverter). The converter contains Insulated Gate Bipolar Transistor (IGBT) and produces harmonic currents on the grid. To reduce the effect of the harmonics, capacitors and Electro Magnetic Compatibility (EMC) filters are installed which reduce the high-frequency currents.

5.4.2 Design of the ENERCON 900-KW Wind Turbine

The ENERCON E-44 is also of the DFIG type. The specification for the ENERCON 900-kW turbine is shown in Table 5.2.

Table 5.2 ENERCON 900-kW wind turbine specifications

ENERCON E-44/900-kW	
Rated power	900-kW
Rotor diameter	44 m
Hub height	45 m/55 m
Turbine concept	Gearless, variable speed, single blade adjustment
Rotor type	Upwind rotor with active pitch control
No. of blades	3
Swept area	1521 m^2
Rotational speed	Variable 12–34 RPM
Generator	ENERCON direct-drive annular generator
Grid feeding	ENERCON inverter
Cut-out wind speed	28–34 m/s
Remote monitoring	ENERCON SCADA

5.4.3 Electricity Market

The Single Energy Market (SEM) is the wholesale market for the island of Ireland, regulated jointly by the Commission for Energy Regulation and its counterpart in Belfast, the Utility Regulator (http://www.uregni.gov.uk). The SEM is structured as a compulsory pool market with capacity payments. Within the market, all price making generators must bid their short-run costs into the pool. Electricity suppliers purchase electricity from the pool to cover their consumer's demand for each half-hour period throughout the day. Once generators have submitted their bids to the Single Electricity Market Operator (SEMO), an initial software run is conducted to determine a Market Schedule which forecasts the System Marginal Price (SMP) for each half-hour trading period. The SMP, calculated by the Market Scheduling and Pricing (MSP) software is set by the most expensive generator required to meet supplier demand in a half-hour trading period. All generators which produce electricity in a trading period receive the SEM pool price for that period, which for most generators is higher than their short-term cost of producing electricity. As wind generators do not consume fuel, they have no short-term costs and hence can bid a zero price to the SEM. As price takers in the SEM, they receive the SMP set by the most expensive generator for their output in that half-hour trading period.

5.4.4 Renewable Energy Generation Incentives

As part of Irelands' efforts to achieve 40% of electrical energy produced from renewable sources by 2020, the government has incentivised the production of renewable electricity to encourage entrepreneurs to invest in renewable energy generation. One such support method is the REFIT 2 Renewable Energy Feed-In-Tariff support scheme (DCENR 2015). The cost of the REFIT support scheme is covered by a Public Service Obligation (PSO) levy, which also promotes other market interventions such as peat generation and the provision of peaking plant. These PSO costs are levied on all customers. Any renewable generator that wants to benefit from the government REFIT subsidiary scheme needs to have a Power Purchase Agreement (PPA) with a licensed supplier. The REFIT 2 scheme operates by guaranteeing a minimum price for new renewable generators for electrical energy exported to the grid for a period of fifteen years. A base price per MWh of €66.35 was agreed for onshore wind projects above 5-MW and €68.68 below 5-MW, these prices are index-linked to the Consumer Price Index (CPI) on an upward only basis. In addition to this, a balancing payment of €9.90 per MWh is made to the supplier for any electricity exported onto the grid; this balancing payment is not index-linked. If the market price is equal to or greater than the sum of both the base price and the balancing payment no REFIT is payable to the supplier, and if the market price is less than the sum of both, the REFIT payment is calculated as the difference between the two. The main difference between REFIT 1 and REFIT 2 is that the balancing payment is not paid out if the market price is above the sum of both the base payment and the balancing payment in REFIT 2, whereas in REFIT 1 a balancing payment of 15% of the value of the base payment is paid out even if the market price is above the sum of the base and balancing payments combined. A policy document by Doherty and O'Malley (2011) highlighted some inefficiency in the Irish REFIT scheme. Research by Boomsma et al. (2012) assessed two of the most extensively employed renewable energy support schemes operating in Norway, namely, feed-in-tariffs and renewable energy certificate trading. The authors carried out a Nordic case study based on wind power and found that the feed-in-tariff encourages earlier investment in the wind industry, but renewable energy certificate trading creates incentives for larger projects. The REFIT scheme, supporting the development of renewable electricity since 2009, has provided certainty to Irish renewable energy projects by guaranteeing a minimum price for the energy produced. The scheme was closed on 31st December 2015. The scheme that is replacing REFIT is entitled the 'Renewable Energy Support Scheme' (RESS) and is also supported by the Public Service Obligation levy. The main difference in the two schemes is the shift from a guaranteed fixed price for renewable generators (REFIT) to a more market-oriented (auctions) where competitive bidding between renewable generators determines the support (RESS). Any companies contracted under the REFIT2 scheme before the window closed are continuing to get a guaranteed minimum price for the electricity generated for 15 years.

5.4.5   Business Expansion Scheme (BES)

This scheme, superseded by the Employment and Investment Incentive Scheme (EIIS), is an investment incentive scheme whereby relief from income tax is available by way of a reduction from income to individuals who invest long-term risk capital in ordinary shares of unquoted SME companies resident in Ireland (Davy and BDO 2014). The scheme allows an individual investor to obtain income tax relief on investments up to a maximum of €150,000 per annum in each tax year up to 2020. Relief is initially available to an individual at up to 30%. Up to a further 11% tax relief is available where it has been proven that employment levels have increased at the company at the end of the holding period or where evidence is provided that the company used the capital raised for expenditure on research and development (http://www.revenue.ie). This BES scheme is used to partly finance the wind farm project in this research.

Table 5.3 Capital cost of the 3.5-MW wind turbine project

Capital cost (Per MW)	€1,150,000
ESB connection cost (Per MW)	€200,000
Civil works	€450,000
Roads contribution	€70,000
Initial development costs	€157,350
Total capital cost of project	€5,402,350

5.5 Results and Discussion

5.5.1 Cost of 3.5-MW Wind Farm Project

There is no promoter equity given for the cost of the project. It is fully funded by the two owners, with financial assistance from the Business Expansion Scheme (BES) which provided €1,840,250 in capital at the beginning of the project. The money is returned in year 5 with an extra 10% on redemption. Most of the remaining cost, €3,565,551 is provided by a long-term, 15-year, bank loan. The loan mechanism process is an amortised type whereby the owners make regular repayments which include both interest and principal amounts and in doing so reduces the amount of money owed, principle, on the loan over time. At the beginning of the loan repayments, the interest and principal monetary values are approximately equal, i.e. €170,000 per annum. The interest repayments subsequently decrease as more and more of the principal is paid off over time. There is no corporation tax to be paid on the investment for the first eight years. The capital costs include turbines, unit transformer, crane, non-buoyant foundation, protection equipment, grid-code compliance devices, and testing (Table 5.3).

5.5.2 Funding for 3.5-MW Wind Farm Project

Table 5.4.

Table 5.4 Total funding for 3.5-MW wind farm

Business expansion scheme (BES)	€1,840,250
Monies provided by project owners	€157,350
Bank loan	€3,565,551
Tax exemption	−€160,801
Total finance	€5,402,350

Table 5.5 Annual revenue generated for 3.5-MW wind farm

Turbine output (kW)	3500
Hours in day	24
Days in year	365
Turbine availability	0.97
Turbine efficiency	0.97
Capacity factor	0.34
Price per kWh	€0.08
Yearly revenue	€804,282

5.5.3 Annual Revenue Generated from 3.5-MW Wind Farm

The turbine produces revenue of €804,282 per annum, as shown in Table 5.5. This value is based on a unit (kWh) price of €0.082. This unit price is not a fixed rate and may either increase or decrease depending on supply/demand market conditions and government support schemes (REFIT or RESS). Whereas a sensitivity analysis is not carried out in this research, it is worth noting that the revenue generated from such a wind farm development is likely to fluctuate on a daily basis.

5.5.4 Financial Analysis on 3.5-MW Wind Farm

The four turbines in the wind farm arrangement produce a combined electrical energy output of 9,808,318 kWh units of energy annually. The energy output value is calculated based on monthly energy statements. The monthly energy statements use digitally recorded generated kWh unit measurements, with the digital meters installed in the switch-room shown in Fig. 5.2. The energy transcripts are accessed by the researcher from the wind farm owners. The annual electrical energy output is used in the following financial analysis of the investment decision.

Net Present Value (NPV); the NPV investment appraisal method considers all of the costs and benefits of the turbine installation and makes a logical allowance for the timing of these costs and benefits. The time factor is a crucial factor as the investors do not see €10,000 received now as equivalent in value to €10,000 receivable in a year. The three reasons for this are (i) Interest lost; (ii) Risk; (iii) Effects of Inflation. The NPV method makes a direct comparison between the sum of the inflows over time and the immediate investment. The cash benefits over time are discounted, depending on the interest rate and the period (year) in which the benefits arise. The discount factor is 6% in this case. The Net Present Value (NPV) for the project is calculated as €2,043,752. The NPV summates all the costs and benefits over the twenty years of the lifetime of the project. When the NPV value is positive, it indicates that the risks associated with the investment are worth taking.

Fig. 5.5 SP & DCF (SP = Simple payback, DCF = Discounted cash flow) for 3.5-MW wind farm

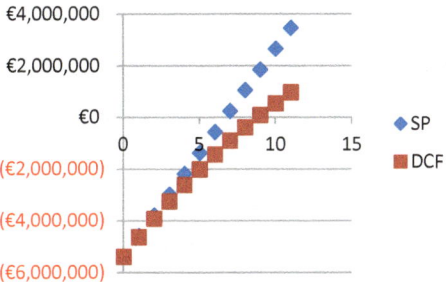

The Payback Period (PP, also called Simple Payback, SP data in Fig. 5.5) of the project is calculated as (€5,402,350)/(€804,282) = 6.7 years. The PP is the length of time it takes for the initial investment to be repaid out of the net cash inflows from the turbine installation. It is a much-improved result compared to the micro-generation wind turbine results Kealy (2014) discussed previously, which had a Payback Period (Simple Payback) of 23 years. The advantages of the PP method are that it is quick and easy to calculate and is easily understood by the person making the investment decision. If the financial benefits of the investment are discounted, due to the time factor as discussed in the previous chapter, then the Payback Period is between eight years and nine years, as demonstrated in Fig. 5.5 (DCF, Discounted Cash Flow data points). This simple payback calculation is slightly inferior to the 5.2 years calculated by Tran and Chen (2013) on a rural area of Vietnam. However, the hub height in the research by Tran and Chen (2013) was 80 m high, significantly higher than the height of the turbines assessed in this case study.

5.6 Conclusions

Based on the operational performance of the four-turbine 3.5-MW wind farm in producing electrical energy kWh units to feed into the National Grid, the results of this case study show that this was a profitable investment decision by the two leading investors. While the capacity factor did not reach the pre-connection predicted value (40.29% for Phase 1 and 36.24% for Phase 2, Sect. 5.3.2), it did reach a satisfactory level of 34% (Sect. 5.3.4). It was a significant improvement on the 10-kW small-scale wind turbine discussed by Kealy (2014). There may be several factors affecting this result. There is a highly efficient control system on each of these large turbines. This control system allows the system to maximise its output, even in low wind conditions. The Vestas Converter System (VCS) ensures a steady and stable electric power from the turbine. The stability is achieved by regulating the pitch of the blades and controlling the current in the rotor circuit of the generator. The significant number of kWh units produced by the carbon-neutral 3.5-MW wind farm (9,808,318 kWhs annually) is expected to *offset* the number of kWh units that must be provided by traditional means, i.e. a mix of fossil fuels and renewable sources generally associated with the national electrical grid as stated in Sect. 5.3.2. This offsetting approach benefits the environment by the reduction in CO_2 emissions due to the burning of fossil fuels in the traditional electricity generation process mix. However, to add rigour to the findings of this study, two further investigations are needed (i) the energy benchmarks should be employed to validate the equivalent reduction in kWh units supplied through 38-kV Supply A and 38-kV Supply B power lines in Fig. 5.3 and (ii) the 3.5-MW wind farm power quality should be evaluated.

There were extensive preliminary on-site tests carried out before the project went ahead. These involved an environmental impact assessment and also wind speed analysis on the site. These tests ensured the suitability of the site, gaining prior knowledge that it is noted as a 'good' site, and helped to minimise the risks associated with the investment.

The design, installation, and commissioning engineers were experts in the wind industry, and this manifested itself in all aspects of the project. The risk associated with the investment for the wind farm installation in this research appears to be a risk worth taking. Financial results indicate a healthy bank balance and good income is achieved from the four turbines in the wind farm project. A simple PP of less than seven years is acceptable to the project owners. To confirm this statement, the wind farm owners are in the planning stage for a new 3-MW single turbine close to the existing site used in this research.

While the financial (economic) results are favourable for the case study presented in this chapter, further investigations are needed before the project can be definitively confirmed as contributing positively to the TBL understanding of sustainable development.

References

Abeliotis, K., & Pactiti, D. (2014). Assessment of the environmental impacts of a wind farm in central Greece during its life cycle. *International Journal of Renewable Energy Research, 4*(3), 580–585.

Boomsma, T. K., Meade, N., & Fleten, S. E. (2012). Renewable energy investments under different support schemes: A real options approach. *European Journal of Operational Research, 220*, 225–237.

Catalin, C., & Nicoleta, R. (2011). International biomass trade and sustainable development: An overview. *Annals of the University of Oredea, Economic Science Series, 20*(2), 47–54.

Davy & BDO. (2014, December). The Davy EII tax relief fund 2014.

DCENR. (2015). Department of Communications, Energy and Natural Resources. http://www.dcenr.gov.ie. Accessed on 12 March 2015.

Department of Communications, Energy and Natural Resources, REFIT Scheme. http://www.dcenr.gov.ie/Energy/Sustainable+and+Renewable+Energy+Division/REFIT.htm. Accessed on 26 March 2015.

Doha. (2012). Doha Amendment to the Kyoto Protocol. http://unfccc.int/kyoto_protocol/doha_amendment/items/7362.php. Accessed on 20 October 2014.

Doherty, R., & O'Malley, M. (2011). The efficiency of Irelands renewable-energy-feed-in-tariff (REFIT) for wind energy. *Energy Policy, 39*, 4911–4919.

Henaghan, D. (2013). Analysis of the capacity factor of Irish wind energy, Dissertation as part of the MSc in Energy Management (DT711), Dublin Institute of Technology.

Iyasere, E., Salah, M. H., Dawson, D. M., Wagner, J. R., & Tatlicioglu, E. (2012, May). Robust non-linear control strategy to maximise energy capture in a variable speed wind turbine with an internal induction generator. *Journal of Control Theory and Applications*, 10(2), 184–194.

Kealy, T. (2014, April). Financial appraisal of a small scale wind turbine with a case study in Ireland. *Journal of Energy and Power Engineering*, 8(4), 620–627. https://doi.org/10.17265/1934-8975/2014.04.004.

Kose, F., Aksoy, M. H., & Ozgoren, M. (2014). An assessment of wind energy potential to meet electricity demand and economic feasibility in Konya, Turkey. *International Journal of Green Energy, 11*(6), 559–576.

Kyoto. (1997). Kyoto protocol, United Nations framework convention on climate change. http://unfccc.int/kyoto_protocol/items/2830.php. Accessed on 20 October 2014.

Tran, V.-T., & Chen, T.-H. (2013). Assessing the wind energy for rural areas of Vietnam. *International Journal of Renewable Energy Research, 3*(3), 523–528.

Sustainable Energy Authority of Ireland. (2013). Renewable energy policy. http://www.seai.ie/Renewables/Renewable_Energy_Policy. Accessed on 12 March 2015.

Chapter 6
Stakeholder Outcomes in a Wind Turbine Investment: Is the Irish Energy Policy Effective in Reducing GHG Emissions and Electricity Costs by Promoting Small-Scale Embedded Turbines in SMEs?

Abstract As a member state of the European Union (EU), Ireland has adopted an energy policy which includes promoting wind-powered electricity generators as an economically viable, GHG-reducing alternative to environmentally damaging fossil-fuel-driven electrical generators. This longitudinal, inductive in-depth study investigates the outcomes for the government, investors, and other stakeholders involved in a 300-kW wind turbine project investment by a Small–Medium Enterprise (SME) based in rural Ireland. A case study/action research methodology is used to acquire and analyse quantitative numerical data from multiple sources, including electrical power and energy meters, historical electricity bills and company sustainability reports. Numerous site visits were organised, where the researcher familiarised himself with the culture and experiences of the employees in the participant company. The study found that the installation of the 300-kW wind turbine did not contribute significantly to the EU-binding Green-House-Gas (GHG) national emission reduction targets and had minimal positive effects on the electricity costs for the business. Indeed, the turbine appears to have significant adverse effects such as a need for an increased Maximum Import Capacity and deterioration in the utility power factor. Empirical measurements on the wind turbine output identified a constantly ramping power output signal. The findings also serve to question the effectiveness of the sustainability reporting framework. The number of energy units produced by the turbine was overstated in the SME's sustainability report mainly due to undetected erroneous energy readings. Caution should be exercised when business owners select alternative energy providers who claim to be experts in the energy field but may have limited knowledge in this area of wind energy. This exploratory study is of benefit to all stakeholders, including the national government who are promoting wind energy as a significant player in the overall energy policy as they target a reduction of Green-House-Gas emissions.

Keywords Wind turbine · Sustainability reporting · Green-House-Gas emissions · Business ethics · Embedded generation · Power output

© Springer Nature Switzerland AG 2020   115
T. Kealy, *Evaluating Sustainable Development and Corporate Social Responsibility Projects*, https://doi.org/10.1007/978-3-030-38673-3_6

6.1   Introduction

The somewhat recent phenomenon of globalisation has helped to create interconnectivity and interdependence between people, businesses, cultures, and nations as never before. This phenomenon has introduced a new kind of responsibility where terms like Corporate Social Responsibility (CSR), Corporate Responsibility (CR), and Corporate Citizenship (CC) have emerged (Aspling 2013). Corporations surely wish to see a developed society so that society continues to be their paying customers. It is in this context that the concept of sustainable business development has emerged as an alternative philosophy to the widely used growth-driven and profit-driven, neoclassical economic theory which has put enormous pressure on our vast but finite natural resources, causing environmental degradation which has the potential to threaten our wealth and even our existence (Catalin and Nicoleta 2011). The supply of oil, coal and gas would appear to be approaching the end period of their life cycle (Shafiee and Topal 2009). It is argued by many scientists that the burning of fossil fuels in power generating plants is contributing to environmental degradation, which over time could jeopardise our wealth and even our existence (Catalin and Nicoleta 2011). In recent times research papers, reports and international conferences such as Kyoto (1997), Doha (2012), and Rio (2012) have drawn attention to an imminent environmental, economical, and human crisis. The wind energy industry grew out of this emphasis on sustainable business development, as wind turbines were seen as an environmentally friendly alternative to fossil-fuel burning plants for electrical power generation. This current research study investigates in detail the operational stage of one such wind energy project undertaken by a Small–Medium Enterprise (SME) based in rural Ireland. Ireland is committed to reducing its dependence on fossil fuels in line with European Union directives (Department of Communications, Energy and Natural Resources, DCENR 2014). The Irish government has initiated a policy whereby 40% of its electricity is intended to be generated by renewable sources by the year 2020 (SEAI 2013). Of this figure, it is envisaged that a significant portion is expected to come from wind energy which has prompted increased interest in wind farm development.

This research is carried out under the stakeholder theoretical framework. Stakeholder theory holds that business activity affects not only the people who have capital invested in the business and desire a profitable return on that investment, but its activities also have an effect on additional actors including its employees, customers, competitors, suppliers, government, and media. Stakeholder theory is an integral part of Corporate Social Responsibility (CSR) as claimed by Agudo-Valiente et al. (2015). Holme and Watts (2000) define CSR as a continuing commitment by businesses to economic development while improving the quality of life of the workforce and their families as well as of the local community and society at large. Stakeholder theory is further developed for a long-term perspective in Sustainable Development (Krisnawati et al. 2014). Research by Hansen et al. (2014) found that sustainability management and CSR have become tightly integrated. Sustainability management in businesses must account for *financial, environmental,* and *human* issues. Elkington (1997) identifies these discrete elements as the three-P's (Profit, People, and Planet). Kealy (2014) claims that these separate components are interconnected; if one part is underperforming, then it can hinder the sustainable development of the entire business. Kuhnen and Hahn (2019) concur with this sentiment by claiming that environmental protection and economic prosperity are a means to ensure the social dimension of sustainability is observed. This research focuses on one specific aspect of CSR, namely the environment, and more concisely one element of the environment, namely climate change. It is alleged by many scientists that the climate is changing at an alarming pace with, among other things, an increase in global temperatures that could have a catastrophic effect on the peoples of the world in the not-too-distant future. It is claimed that the phenomenon of global warming is partly due to the emission of Green-House Gases into the atmosphere by the burning of fossil fuels (gas, oil, and coal) to power electricity generators. It is therefore incumbent on society, with the help of governments, to develop alternative, renewable sources of energy to reduce the dependence on fossil fuels. It is in this context that wind energy generators are promoted as having the potential to contribute positively to the seemingly unsustainable global energy situation. However, there is minimal empirical evidence to suggest that wind energy policy is having a profound effect. This research into a wind turbine installation by an SME as part of their CSR initiatives helps to consider and evaluate the Irish government policy as they aim to reduce binding EU emission targets, which may be very costly if these targets are not achieved.

Types of Wind Energy Systems: One of the problems associated with wind-generated power is the random nature of the power output due to the randomness of the source wind. The output power variations are a function of short-term wind speed variations (Wan 2004). Long duration wind variations cause problems, e.g. congestion and reliability of the system, for the network operators who can put in place established strategies like generation forecasters, dispatch and contingency analysis, and real-time control. Long duration wind variations are generally in the hour or half-hour time frames. By contrast, turbulence and gusts are examples of short-duration wind variations, variations in the wind velocity from 10-min averages, the effects of which are more challenging to observe. Research by Roy (2013) sought to estimate the standard deviation (SD) of power output by a wind turbine generator in the presence of short duration wind variations. The author uses the two-parameter Weibull statistics as a description of wind variations around stable mean values. The Weibull probability distribution function uses the quadratic relationship between wind power and wind speed. The short duration output power and the output power variability are estimated using the generally available turbulence intensity measure. Turbulence intensity is the ratio of the wind speed standard deviation to the mean wind speed, determined from the same set of measured data samples and taken over a specified period (I.S. EN 61400-1: 2005). Research by Lin et al. (2012) sought to quantify the variations in wind turbine power output using field-measured wind power output data, the data being obtained using 1-min average data. Four different types of wind farms were analysed in the research by Lin et al. (2012):

- Type 1: Induction generator with fixed rotor resistance, fixed speed
- Type 2: Induction generators with variable rotor resistance, variable speed
- Type 3: Double-Fed Induction Generators (DFIG)
- Type 4: Generators with full power converters.

Research by Boutsika and Santoso (2013) concludes by stating that the DGIG type of wind turbine (Type 3) has room for improvement when assessing wind power variability, and a full-power converter (Type 4) may be a better option.

6.1.1 Research Site for 300-kW Wind Turbine

This research was conducted on an SME facility situated in rural Ireland in the northeast of the country. The company is one of the largest growers, packers, and distributors of fresh produce to the retail market. The full-time workforce is 200 people and has grown steadily since its inception in 1982. The company embraced supply-side management at the end of 2013 by installing an embedded 300-kW on-site wind turbine, Fig. 6.1. The company has an approximate annual electrical

Fig. 6.1 Image of 300-kW DFIG on-site wind turbine (Kealy)

energy usage of 1,200,000 kWh units (2013 values) and has targeted a reduction of this value as a means to reduce their carbon footprint as recommended in the sustainability reporting guidelines. An embedded wind turbine was regarded as a strategic initiative in helping to reach these energy reduction targets. To this end, a 300-kW on-site wind turbine was purchased and installed towards the end of 2013. This study mainly concentrates on the quantity of imported electrical energy units in the year before the turbine installation, i.e. 2013 and compares that value with the quantity of imported electrical energy units after installation, i.e. 2014 and also provides an opportunity for the intensive analysis of many specific details of embedded generation. As well as investing in the turbine installation, the company has also outsourced its energy management function to an Irish-owned external energy management company which specialises in supplying natural gas and electricity to the industrial and commercial market.

Embedded generation is an approach that employs small-scale technologies, like wind turbines, to produce electrical energy close to where the company is using the power, thereby reducing transmission losses. In this case, the turbine is installed on the company site and is connected in parallel with the electric utility supply at the electrical intake in the main switch-room, as shown in the diagram in Fig. 6.2. The marketers of embedded generators claim that they offer many potential benefits over traditional power generators. The list of potential benefits includes lower cost electricity, higher power reliability and security and adverse environmental consequences are minimised. Wind energy electrical generators require back-up (embedded) traditional generators such as gas, coal, and peat to compensate for the intermittent

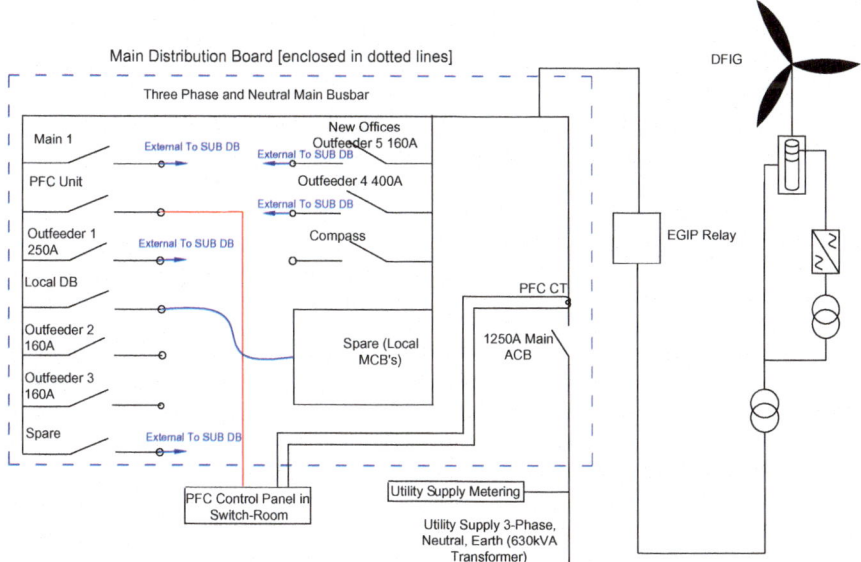

Fig. 6.2 Electrical main distribution board in main switch-room and embedded turbine

nature of wind power generation (Kaffine et al. 2013). Rapid adjustment is required of the fossil-fuel generators in response to increases or decreases in wind generation output.

Cost–benefit analysis is often used in engineering where an effort is made to translate negative and positive utilities into monetary terms. However, in the case of this *new* wind energy technology, cost–benefit analysis is sometimes referred to as risk–benefit analysis because much of the analysis requires *estimating* the probability of certain benefits and costs.

The renewable energy solutions company which supplied the turbine is based in the South-East of Ireland. Access to the turbine tower is via a secure door at its base, accessible only to suitably qualified personnel. The turbine control panel and accessories are all located at the bottom of this tower. A digital controller indicates parameters used in the analysis of this research, which include energy output data, alarm indication data, rotor speed data, and blade speed data. On each site visit, these parameters were noted and recorded.

The three-blade 300-kW Double-Fed Induction Generator (DFIG) turbine (Fig. 6.1) has a rotor diameter of 28 m. It is an asynchronous-type generator with a three-phase 690 V output. An asynchronous (induction) generator is a type of alternating current generator that uses the principles of induction machines to produce power. Asynchronous generators work on the principle that when the rotor blades are rotating faster than the synchronous speed of the wind turbine, the machine is operating in generator mode and produces electrical power. If the rotor blades rotate slower than the synchronous speed of the wind turbine, the machine operates as a motor and draws current from the supply. A synchronous generator usually draws its excitation power from the National Grid (50 Hz in Ireland). The machine cannot start on its own as a generator. The machine starts as a motor. The DFIG speed adjustable design uses an electronic controller to convert a variable frequency alternating current (AC) to direct current (DC) and back to AC at the grid frequency, i.e. 50 Hz. The AC/DC/AC conversion stage uses the thyristor technology in this installation. A thyristor is a solid-state semiconductor device that is used in turbine control as a switch. The thyristor technology has been in use since 1956, and while they can be used on large power switching, they have certain disadvantages and have virtually been replaced by other devices with superior switching characteristics, e.g. Insulated Gate Bipolar Transistors (IGBT) devices. The rated wind speed is 13.5 m/s, and the swept area is 615 m^2. The hub height is 31 m.

The net cost of the project was €280,000 with an expected useful life of the turbine of 20 years. The wind turbine supplier predicted that the 300 kW turbine installation would save 50% of the overall electricity imported from the National Grid, i.e. the turbine would provide 600,000 kWh units of electricity per annum. They also predicted a Payback Period (PP) of 5 years. The distance between the wind turbine and the main electrical switch-room is 240 m. There are two cables in parallel between the turbine and the switch-room, each cable is 4-core, 240 mm^2, Steel Wire Armour (SWA) aluminium type. The business tariff used is the Maximum-Demand Low Voltage Quarter-Hour type. Before the installation of the turbine, the Maximum Import Capacity (MIC) was set at 320-kVA. The 'Utility Supply Metering' arrangement (Fig. 6.2) contains an import/export facility whereby any units of energy produced by the turbine and not instantaneously required on-site can be exported back to the National Grid for a kWh unit price agreed between the turbine owner and the electrical utility provider. A total of 6332 kWh units of electricity were exported in 2014. The remaining kWh units are assumed to be used on-site.

6.1.2 Plant Overview for 300-KW Wind Turbine

The turbine under test is a Type III Double-Fed Induction Generator (DFIG) type installed at the end of 2013. It has a three-phase 690 V output switched to the National Grid through thyristor switching devices. The speed of the generator is 1500 Revolutions/Minute (RPM). Because the turbine supplies the client's load in parallel with the National Grid, an Embedded Generator Interface Protection (EGIP) relay is used to check that all the conditions are met to embed the two supplies safely together. The 300-kW rated turbine produced 460,842 kWh units of electrical energy in 2015, according to the digital meter indicator in Fig. A6.3. This 460,842 kWh value gave an annual capacity factor of 17.53% in 2015. When the SME purchased the turbine and installed it at the end of 2013, the turbine supplier stated that it would deliver half of their electricity units per annum, equating to approximately 600,000 kWh units of electrical energy provided by the wind turbine. The expectation was that these 600,000 carbon-neutral kWh units would displace the same number of units generally supplied by the National Grid, much of which is generated by fossil-fuel burning plants, emitting much GHG into the atmosphere in the process. The schematic diagram of the installation is shown in Fig. 6.2.

The EGIP relay is shown to the right of the schematic diagram in Fig. 6.2. All of the tests are carried out between the EGIP relay and the main distribution board. Renewable energy data from the system shown in Fig. 6.2 is collected by the SEAI. At the beginning of each calendar year the SEAI, under the guise of the Energy Policy Statistical Support Unit, contact businesses with embedded/autoproducing wind turbines and request information on the number of energy units generated by the turbine in the previous year and the size of the turbine in kW. The SME is also asked about the number of units used on-site and the number of remaining units exported to the National Grid. The information is essential for Ireland to meet its international reporting obligations under the European Energy Statistics Regulation of 2008, No. 1099. Regulation (EC) No. 1099/2008 of the European Parliament and the Council establishes a common framework for the production, transmission, evaluation and dissemination of comparable energy statistics in the European Community. This framework relies on suitably qualified personnel in participant SMEs to carry out the task accurately. The Electricity Supply Board (ESB) utility supply meter shown in Fig. 6.2 is set up to measure exported units, as well as imported units, and indicated that 27,828 kWh units of energy were exported in 2015 meaning that the remaining units 433,014 kWh units were used on-site by the SME. The larger non-embedded wind farms are metered at the grid interface by either ESB Networks or Eirgrid where the turbines are providing energy directly into the National Grid. The way in which it is intended to reduce GHG emissions is that for every kWh unit of electrical energy produced by the carbon-neutral wind turbine there is a displacement of a unit of electrical energy which would otherwise be supplied from the National Grid, with a substantial portion of the same currently being derived from fossil fuel. The burning of fossil fuel sufficient to generate one unit of electrical energy (kWh) produces 456.6 grammes of CO_2 (SEAI 2015, p. 82) which is emitted into the atmosphere (Table A4.1, 2014 value).

6.2 Methodology

The case study/action research methodologies were chosen on this singular business entity because it was considered that it allowed the researcher to investigate the case in depth and within its real-world context (Yin 2003). Case study and action research methodologies are suited to the interpretive paradigm under which this research is carried out (Kivunja and Kuyini 2017). The interpretive paradigm makes use of the researchers' technical and business background/experience (Electrician, B. Eng. in Electronic Engineering, and PG Dip in Management and Marketing, Ph.D) to interpret the data that is collected and analysed in the study. The single business case study/action research is also selected because it was deemed that it reflects many socially responsible SMEs in Ireland and it provides the researcher with an opportunity to observe and analyse the first-hand experiences on stakeholders of investing in a 300-kW on-site wind turbine. The SME upon which this research is based on had made decisions to participate in sustainability initiatives, and as part of their CSR efforts have installed a 300-kW wind turbine at a capital cost of €280,000. Quantitative methods are used to collect and analyse the numerical data over a protracted period. The resulting data is presented using tables and graphs and statistically analysed to describe the central tendency and the dispersion of the data. The Microsoft Excel and IBM SPSS Statistics software are used for data analysis. Data was collected from multiple sources to ensure rigour to the case study. This data includes company sustainability reports, electricity utility bills, digital meter indicators and analogue (rotating disc) meter indicators. The evaluation of the evidence from these multiple sources is triangulated to confirm and corroborate the findings. The reliability of the data is assured by measuring the power output from the turbine on several site visits incorporating a range of wind speeds. The digital instrument used to measure power output from the turbine is a new high-quality instrument (Fluke 1735 Power Logger). Measurement validity is assured as the measuring instrument is calibrated and designed to measure power flow and quality on a range of electrical systems. The Power Log V4.3.1 software is used in conjunction with the meter to obtain data for this research. The Fluke meter instantaneously measures and stores voltage, current, and power factor data on connected loads (actual measurements) and outputs a value at half-second intervals. The meter calculates minimum, average and maximum values within the half-second (0.5 s) period. The generated interval data is used to process values for mean, standard deviation and ramping (or rate-of-change) of wind turbine output power. The power factor is measured using a zero-crossing analysis of the voltage signal compared with the zero-crossing of the current signal. The power and energy values are subsequently calculated using these measured voltages, current and power factor values. When the measured power factor indicates a negative value, this indicates that power flow is from the installation (utility) to the turbine. When the power factor value is positive, this indicated that power flow is from the turbine to the installation. The power directional arrows printed on the current transformer (CT) clamps point *from* the turbine *to* the installation. The discrete data is recorded at half-second intervals where the Fluke meter has a storage capacity of 36 min, giving a total number of 4320 individual cases. Data is transferred from the Fluke meter to the researcher's PC via a USB link. The Fluke digital meter is set to record a sample every half-second (the minimum setting permitted by the meter) as this gives an accurate indication of the amount of ramping (short-term variations) in the turbine power output signal. In addition to being interested in the mean (average) value for the power output from the turbine, this research is very interested in the amount of variation shown by the distribution, i.e. the extent to which the data values (Watts) are spread around the mean. This phenomenon is called '*dispersion*', and it is a measure of how widely spread a distribution is. While two different data sets may exhibit the same mean value, how widely spread out they are might be different. The most commonly used method of expressing the diversity of a data set is the '*standard deviation*'. The standard deviation (SD) calculates the average amount of deviation from the mean. The standard deviation reflects the degree to which the values in a distribution differs from the arithmetic mean. It is usual practice to present the standard deviation at the same time as the mean since it is difficult to determine the meaning in the absence of the mean. The coefficient of variation (COV) metric is also calculated on the 300-kW turbine power output data. The COV is a measure of the dispersion of a probability distribution or frequency distribution and is defined as the ratio of the standard deviation to the mean. Reliability of the analogue Rotating Disc Meter (RDM) was tested before and after the instrument was connected to the turbine. The RDM was connected to an electrical system with a sealed calibrated fixed utility meter (Fig. A6.1) and the results compared. The RDM consistently showed a 98% accuracy rate compared to the Electricity Supply Board (ESB) reading.

Access to the Research Site: Full access to all areas was permitted to the researcher during the data collection stages. Many site visits were carried out over a protracted period. On each visit, energy meter readings from the digital meter and the electromechanical rotating disc meter were noted and recorded while on five site visits relevant data was collected using the Fluke Power Logger. The researcher considered that the case study/action research approach would allow for the exploration of this relatively new concept of a singular embedded wind energy system where uncertainty still surrounds the benefits of such a project to business (Gray 2009). This relatively new phenomenon of the adaptation of wind energy systems by single business entities has led to a plethora of new 'energy companies' which are providing this service. However, the effectiveness of these projects concerning cost savings for the companies which undertake such energy systems is yet to be proven. The researcher's background as a qualified electrician coupled with level 9 and level 10 qualifications in Engineering, Management and Sustainability contributed to a comprehensive data collection and analysis of appropriate variables in this study. The researchers' interaction with the case study participants in their real-life environment afforded an ethnographic method of inquiry. This method included ongoing interaction with the business owner, site operations manager, company sustainability manager, wind turbine vendor, energy manager (external contractor), and electricity utility provider. As is the case with ethnographic studies, care was taken to minimise imposing the researchers' implicit perspectives on answers to participants' views on the wind turbine project (Thuita et al. 2019).

The researcher believed that due to limited detailed figures/research being available on singular embedded wind turbine systems, a case study/ethnographic approach at the outset would provide a basis for further studies on this phenomenon. To this end, as much detail and information and recording as possible were gathered to ensure the robustness of this case study (Yin 2003). The theoretical stance that was taken at the outset within this study by the researcher was one that seeks to evaluate this wind energy project, the theoretical research hypothesis being that the adaptation of this renewable energy system would benefit the company financially. In doing so, however, the researcher was cognisant of the fact that the findings of a singular project such as this might be interesting and phenomenal in itself, but caution would need to be taken not to generalise the results to the larger population of other such projects.

6.2.1 Data Collection for the Evaluation of the 300-kW Turbine Investment Decision

This study involved analysing the effectiveness of this wind turbine installation in reducing the cost of the electrical energy bill. From the outset, it was clear that singular source data collection, i.e. just analysing the electricity bills would not be sufficiently detailed to effect robust findings. It was, therefore, necessary to collect data from several sources over four years, 2011–2014. A specific focus was then concentrated on the values for 2013 against 2014 to make accurate claims. The data collection was painstaking and detailed, collected manually and recorded by pen and paper and then transcribed onto spreadsheet software. This process demanded weekly visits to the installation for the years 2011–2014. The primary sources of measurement data were

- **Utility Electrical Bills**

  Number of imported energy units (kWh) per monthly/yearly period and
  Total bill cost per monthly/annual period in Euro.

- **Turbine Energy Output**

  Data recorded in kWh from digital energy meters in the turbine tower yearly.

- **Wind Data from Met Eireann**

Wind speed measurements (in knots per second converted to meters per second) for the four years 2011–2014 from three wind recording sites local to the case study installation, namely Mullingar, Dunsany, and Dublin Airport. Average daily and monthly wind speeds were recorded, and any adverse weather noted. Data were then transcribed to Excel spreadsheets and analysed.

- **Factory Production Output**

Production output in tonnes per year for 2011–2014.

Data for this study were accessed during the numerous site visits by the researcher. Full access was given to the researcher by the participant company. This concession included access to the turbine tower and the main electrical switch-room. All data was manually collected. During 2014 the wind turbine was switched and locked OFF for twelve days. This switch off was to allow essential maintenance and testing to be carried out on the wind turbine installation. Production at the factory remained at normal levels during this period. Daily imported kWh units from the National Grid were recorded during this 'OFF' period. When the turbine was switched back 'ON', the daily imported kWh units from the National Grid were recorded for a corresponding twelve days. Both sets of readings corresponded to the same times and days within the twelve-day test period. These interesting and unexpected results are reported in Sect. 6.4.2.1.

6.2.2 Data Analysis for the Evaluation of the 300-kW Turbine Investment Decision

Data analysis in the case study methodology can sometimes prove challenging (Yin 2003). In this piece of research, a primarily time-series analytical approach to the data analysis was utilised. The initial theoretical stance that the turbine would contribute to a reduction in energy costs for the company formed the basis for a time-series analysis of the data. The vast quantity of data, collected in this study for the period 2011–2016, was subsequently analysed and patterns/relationships between the aforementioned key variables described in Sect. 6.3.1 noted and recorded.

6.2.3 Validity and Reliability

Validity and reliability are of particular importance in the case study and ethnographic approaches in research because of the reliance on data generated from limited situations (Gray 2009). This concern indeed is the case also in this study which, although detailed, was confined to a singular alternative energy unit only. The researcher was conscious at the outset of the difficulties in ensuring a logical approach to this research subject, given that the generally accepted hypothesis concerning wind and alternative energy sources was that they were to be embraced as environmentally friendly cost-saving 'green' initiatives. The researcher, therefore, investigated and recorded all possible variables that were or were not impacting on the effectiveness of the turbine installation, and allowed the data collected and the results to speak for themselves.

Reliability was assured by the conscientious documentation of the data over a protracted period and a thorough, structured investigation of significant variables that may have affected this data. The strict field procedures were adhered to every week with easy access to the study site secured, and assistance and interaction of relevant workers. As a qualified electrician/electronic engineer, the researcher complied with all the health and safety guidelines on the site. Time scales contingency plans were identified from the outset.

6.3 Technical Assessment of 300-kW Wind Turbine Project

6.3.1 Energy Benchmarks

The company in this research report to the Bord Bia 'Origin Green' reporting framework as part of their CSR strategy. They signed up for the sustainability programme, which helps them to voluntarily achieve measurable sustainability targets—reducing environmental impact, serving local communities more effectively and protecting the local natural resources (Origin Green 2017). The plans are verified by an independent third-party agency and monitored on an annual basis. Some of the initiatives recommended by the sustainability charter include the investment in renewable technology and the reduction in GHG emissions. It was predicted that the installation of the wind turbine at the end of 2013 would tick the 'renewable energy' and 'GHG emissions' boxes. Table 6.1 indicates benchmarks for the number of kWh energy imported units compared with the number of tonnes of products produced in the factory.

Table 6.1 shows the Origin Green benchmark for 2011, 2012, 2013, 2014, 2015 and 2016. From the figures in Table 6.1, it is clear that the business continues to grow, in terms of tonnage product output. However, the number of kWh energy units also

Table 6.1 kWh per tonne of product sustainability reporting benchmark for SME

Year	Tonnes output	kWh import	kWh/tonne
2011	44,301	723,160	16.32
2012	46,157	921,578	19.97
2013	61,723	1,226,945	19.87
2014	72,307	1,377,185	19.04
2015	71,789	1,500,588	20.90
2016	80,568	1,644,252	20.41

Table 6.2 Increase in factory production and factory energy usage for SME

Year on Year	Production (%)	Electric energy (%)
2011–2012	Increase by 4	Increase by 26
2012–2013	Increase by 33	Increase by 34
2013–2014	Increase by 10	Increase by 12

continued to rise during this period. There is no evidence to suggest that the SME is '*saving*' importing 460,842 kWh units of electrical energy, as stated in Sect. 6.1.2 of this book. Of this value, 27,828 kWh units of energy were exported to the National Grid, and the remaining number of renewable-generated units (433,014 kWh units) was alleged to be used on-site by the connected equipment (Table 6.2).

The figures shown in Table 6.1 indicate an increase in production between 2011 and 2016 with a corresponding increase in the number of electrical energy kWh units imported from the National Grid. The company paid a total of €206,310 for imported electrical energy in 2013. They paid a total of €225,497 for importing electrical energy in 2014, a 9.3% increase on 2013 figures. The '*Origin Green*' sustainability plan outlines baseline and target data as a means to monitor whether the improvements are impacting as desired. In this plan, target area 2 examines the manufacturing processes to check the kWh energy usage input utilised to produce 1 tonne of product output. As can be seen from these results of the fourth column in Table 6.1, there is only a very slight improvement in 2014 compared to 2013 figures. A much more significant improvement would be expected as the wind turbine should be contributing some energy units, therefore bringing down the kWh/tonne of product benchmark value.

The turbine produced an annual (2014) output of 288,025 kWh units of electrical energy. This 288,025 kWh value is significantly less than that which the turbine supplier predicted, i.e. an annual energy output of 600,000 kWh units of electrical energy per annum. The Capacity Factor (CF) of this 300-kW wind turbine is therefore calculated at 10.5%. The amount of CO_2 emitted per year due to importing electricity from the National Grid for factory production is shown in Table 6.3.

Table 6.3 CO_2 emissions due to imported electricity from National Grid'

Year	kWh units imported	Tonnes of CO_2 emitted
2011	723,160	330.2
2012	921,578	420.8
2013	1,226,945	560.2
2014	1,377,185	628.8
2015	1,500,588	685.2
2016	1,644,252	750.8

6.3.2   Power Quality of 300-KW DFIG Wind Turbine

The data from the Fluke electric power meter was downloaded to the IBM SPSS (Statistical Package for Social Science) Statistics software. The SPSS package and the (0.5 s) interval data were used to generate the following graphical representation of the data (Figs. 6.3, 6.4, 6.5, and 6.6). On five site visits on five different days (and under various wind conditions), data was gathered during 36 min and the analytical results of these tests are summarised in Table 6.4.

The coefficient of variation (COV) values for the 300-kW wind turbine power output signals (Figs. 6.3, 6.4, 6.5, and 6.6) are calculated and stated in Table 6.4. The tests were carried out on four different days under different wind conditions. The coefficient of variation method is regularly used in engineering applications and is a valid measure of dispersion (Teoh et al. 2017). The COV is also expressed as a percentage value in this book. The sampling time for the data used to demonstrate graphically the plots shown in Figs. 6.3, 6.4, 6.5, and 6.6 is half a second.

Note that the 'Watts Output' values go below the zero reference line at some stages in the test periods for the data presented in Figs. 6.3, 6.4, and 6.6. These below-zero values indicate a negative power flow, i.e. power is being transferred *to* the turbine from the electricity grid.

Descriptive statistics using the IBM SPSS software-enabled mathematical calculations to be carried out on the data. Three of the exploratory data analysis methods allowed for the calculation of the central tendency (among which is the *mean* value), the *dispersion* of the data, and the *coefficient of variation*. The arithmetic '*mean*' is a method for measuring the average of a distribution. The *mean* value of the power

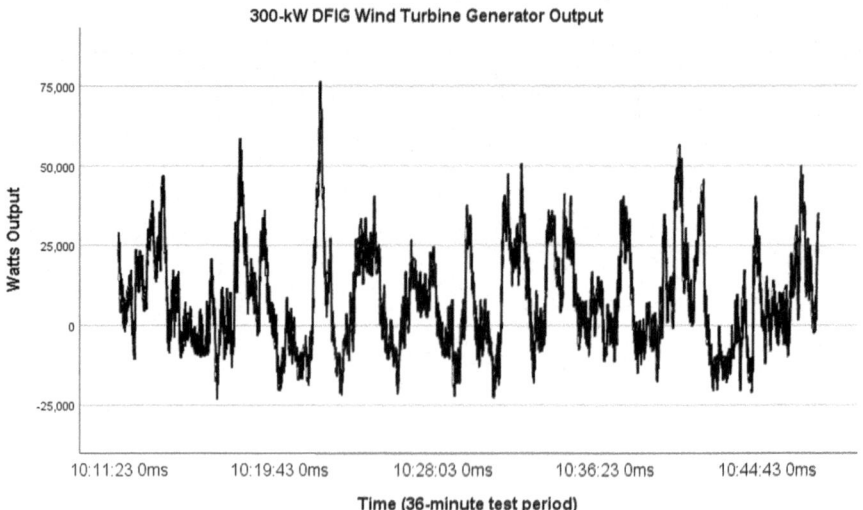

Fig. 6.3 300 kW wind turbine power output on 23 September 2015 with a 10-min average wind speed of 6.4 m/s

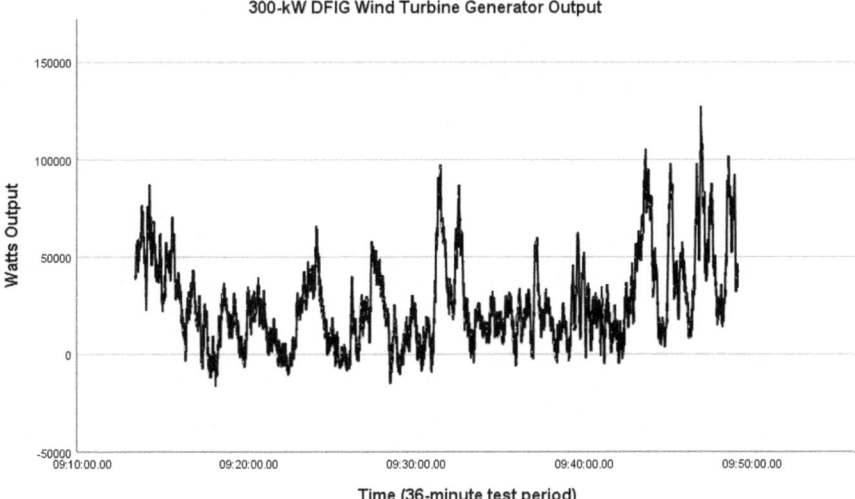

Fig. 6.4 Power output of 300-kW wind turbine on 22 October 2015 with a 10-min average wind speed of 7.3 m/s

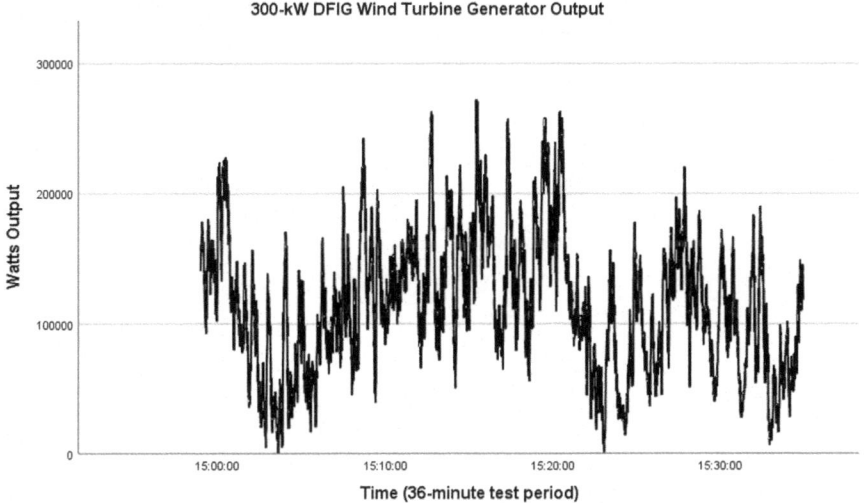

Fig. 6.5 Power output on 10 November 15 with a 10-min average wind speed of 11.4 m/s

output for the 36 min shown in Fig. 6.3 is 7.853-kW (using the SPSS software on the 4319 cases used to generate the graph). The test was carried out during a 36-min test period between 10.11 a.m., and 10.47 a.m. on Wednesday, 23 September 2015. Some of the power output values in Fig. 6.3 are of negative values. These negative values mean that the turbine is acting as an electrical load and consuming power,

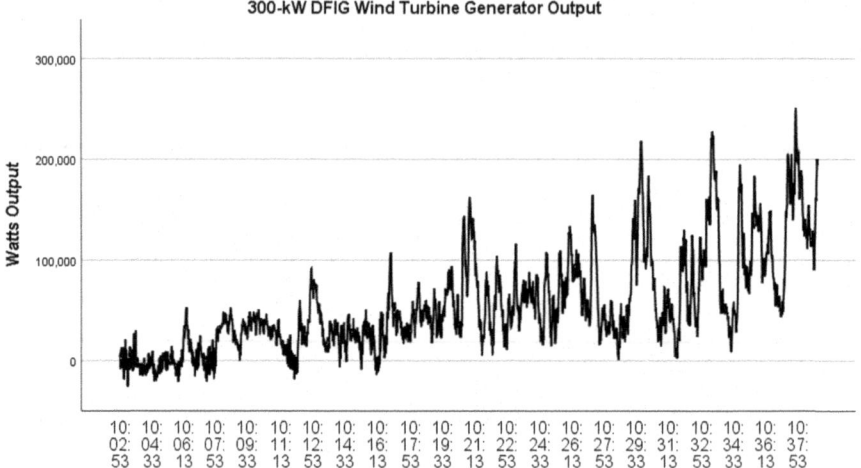

Fig. 6.6 Power output of 300-kW wind turbine on 18 November 2015. The wind speed increased rapidly during the 36-min test period

Table 6.4 Statistical analysis of 300-kW wind turbine power output on five different days

Date	10-Min average wind speed (m/s)	Average output power (36-min) (kW)	Minimum output power (36-min) (kW)	Maximum output power (36-min) (kW)	Standard deviation (kW)	Coefficient of variation
23 September 2015	6.4	7.85	−23.1	76.4	15.8	2.01
22 October 2015	7.3	26	−16.7	126	23	0.88
5 November 2015	2.5	−0.02	−0.9	0.063	0.058	2.90
10 November 2015	11.4	114	0.157	272	52.5	0.46
18 November 2015	12.1	55	−31	254	49	0.89

Table 6.5 Ramping characteristic of 300-kW turbine power output over a one-minute interval period (rate-of-change)

Time	Output power (kW)	Time	Output power (kW)
30:20.500	−2.571	31:20.500	82.862
30:21.000	−5.990	31:21.000	89.417
30:21.500	4.861	31.21.500	84.923
30:22.000	0.878	31.22.000	94.655
30:22.500	4.767	31:22.500	94.561
30:23.000	14.239	31:23.000	94.624
30:23.500	6.900	31:23.500	92.177
30:24.000	7.213	31:24.000	97.446
30:24.500	11.698	31:24.500	86.124
30:25.000	4.202	31:25.000	95.784
30:25.500	7.558	31:25.500	84.525

taking power from the electric utility grid. This imported energy must be paid for and is adding to the cost of the electricity bill for the company.

The standard deviation (SD) for the data shown in Fig. 6.3 is 15.8-kW with an average (mean) power output of 7.85-kW. These values indicate that the average amount of deviation from the 7.85-kW average power output value is 15.8-kW. The *'coefficient of variation'* is a measure of the dispersion of a probability distribution or frequency distribution and is defined as the ratio of the standard deviation to the mean. The *coefficient of variation* for the plot shown in Fig. 6.3 is calculated as 2.01. These values specify a large amount of short-term power output variations and indicate a continuous ramping condition. The mean value for the data shown in the 'Turbine Output in Watts' Fig. 6.4 is 26.3-kW. The duration of the test is 36 min and was carried out on Thursday 22 October 2015 with a 10-min average wind speed of 7.3 m/s. The wattage output power value is changing very rapidly (continuous Ramping) as can be seen from Fig. 6.4. A sample of the source data used to generate Fig. 6.4 is presented in Table 6.5 to calculate the rate-of-change of 300-kW turbine output power measured over a one-minute period. The time between each automatic measurement is 500 ms and the time format for Table 6.5 is MM:SS:msec (Minutes:Seconds:milliseconds). Data in the first two columns are compared to data, one minute later, in the second two columns. The turbine power output changed from 4.767-kW at 30:22.500 to 14.239-kW half a second later. This corresponds to a **rate-of-change** of 9.472-kW per half-second or **18.944-kW per second**. In terms of the rate-of-change of power output over a one-minute period, the power output value at 30:24.000 is 7.213-kW and precisely one minute later at 31:24.000, the power output is 97.446 kW. These values give a **rate-of-change** of the output power of **90.233-kW over a one-minute period**. Other rate-of-change power values appear consistent with these examples. At the beginning of the test period for the data in Fig. 6.6 (10.02 a.m.), the 10-min average wind speed was 6.5 m/s. The wind speed quickly increased, and at the end of the 36-min test period (10.38 a.m.), the 10-min average wind speed had increased to

12.1 m/s. The increase in power output due to the rise in wind speed is demonstrated in Fig. 6.6. A summary of the statistical analysis for each of the five site visits and test data is shown in Table 6.4. This summary is for the data shown graphically in Figs. 6.3, 6.4, 6.5, and 6.6. The low value of average output power in Table 6.4 for the data retrieved on 18 November 2015 (even though the 10-min average wind speed ends up at 12.1 m/s) is because data is averaged over the entire 36-min period. Note from the graph in Fig. 6.6 that the 10-min average wind speed at the beginning of the test is just 6.5 m/s.

6.3.2.1 Power Factor Calculations

The wind turbine installation appeared to coincide with a deterioration in the overall power factor of the plant. This problem necessitated the installation of a new 200-kVAr Power Factor Correction (PFC) unit to be installed in the main electrical switch-room in 2014 (Fig. 6.2). A low power factor, i.e. less than 1, leads to higher currents flowing in circuits causing more significant volt drops which also increased energy losses in cables. A low power factor also contributes to difficulties in electrical switching devices, sometimes causing arcing at the contacts, thereby reducing the lifetime of the switching equipment. In electrical engineering, the power factor of an AC power system is defined as the ratio of the real power (Watts) to the apparent power (VoltAmps) in the circuit. A load with a low power factor draws more current than a load with a high power factor for the same amount of useful power transferred. The excess current is termed 'wattless' current and is not capable of contributing to any work. The low power factor in this case also contributed to the company exceeding its Maximum Import Capacity (MIC) value of 320-kVA on many occasions. In 2014, the MIC value was increased from 320-kVA to 600-kVA by the external energy management company. A new 630-kVA supply transformer was also installed, raised from the previous value of 400-kVA. A sample of the power factor values is given in Table 6.6. Note that the *ideal* power factor is 1.

The power factor values (Table 6.6) indicate values before the turbine was connected (2013), and after the turbine was connected (2014).

Table 6.6 Power factor values for 300-kW wind turbine for comparison pre-connection (2013) and post-connection (2014)

	January	February	March	April
2013	0.92	0.923	0.927	0.916
2014	0.687	0.728	0.86	0.82

6.3.3 Analogue/Digital Energy Measurement

The flexible nature of the inductive approach utilised in this study allowed the researcher to observe patterns associated with the number of electrical energy units produced by the turbine as prescribed by the digital indicator (digital meter in Fig. A6.3). The numbers seemed not to add up. A decision was made to insert a second analogue energy meter in series with the digital indicator to check for accuracy. A three-phase electromechanical rotating disc meter was sourced and is used to carry out a second check on the number of kWh energy units generated by the wind turbine (rotating disc meter in Fig. A6.4) for validation purposes. The rotating disc is acted upon by a set of voltage and current coils, setting up magnetic fields. The interaction between the induced magnetic fields set up a force (driving torque) which is exerted on the disc in proportion to the product of the instantaneous voltage, current and power factor values. A permanent magnet mounted beside the two discs allows for an opposing force to be felt on the disc resulting in the disc rotating at a speed that is proportional to the power, or rate of energy, being used in the load. The number of rotations of the disc is therefore proportional to the energy being consumed by the load in a specified time interval, and the units on the front of the meter indicate kilowatt-hours (kWh). This electromagnetic rotating disc meter is connected just below the Embedded Generation Interface Protection (EGIP) relay (Fig. 6.2) and the kWh units are read and stored during the many site visits and indicated in Table 6.7. The three-phase electromagnetic RDM and connections are shown in Fig. A6.4. Site visits on each of the dates shown in Table 6.7 allowed the comparison of the number of kWh energy units produced by the turbine as indicated by both the Digital Meter (DM) indicator and the rotating disc meter indicator. Ideally, these should be the same value as they measure the same energy signals as shown in Fig. 6.2.

A picture of the digital meter in the turbine tower is shown in Fig. A6.3. The pushbuttons on the front of the panel allow scrolling of the parameters associated with the turbine (in monitoring mode). The digital meter indicator in Fig. A6.3 and the rotating disc meter in Fig. A6.4 are the instruments on which the values in Table 6.7 are based. There is a significant difference between the digital meter reading and the rotating disc meter. The RDM only indicates 59% of the DM value, so the error between the two readings is 41%. The rotating disc meter was checked on several occasions in a controlled setting before and after this case study was carried out for validation purposes and the accuracy of the meter consistently indicating a value of 98%. The validation/test arrangement is shown in Fig. A6.1.

Table 6.7 Comparison of digital meter measurements and analogue rotating disc meter measurements for 300-kW wind turbine

Dates (from date to date)	Digital meter kWh indicated	Rotating disc meter kWh indicated	Rotating disc % of digital value
26 November 2015–28 November 2015	4215	2312	54.9
28 November 15–30 November 2015	5191	2968	57.2
30 November 2015–2 December 2015	7648	4456	58.3
2 December 2015–3 December 2015	87	56	64.4
3 December 2015–7 December 2015	20,210	12,016	59.5
7 December 2015–10 December 2015	13,787	8376	60.8
10 December 2015–15 December 2015	4520	2432	53.8
15 December 2015–21 December 2015	21,427	12,544	58.5
21 December 2015–31 December 2015	22,997	13,400	58.3
31 December 2015–1 January 2016	806	456	56.6
1 January 2016–15 January 2016	8329	4760	57.2
15 January 2016–2 February 2016	58,123	34,968	60.2
Overall	**Total**	**Total**	**Overall**
26 November 2015–2 February 2016	**167,346**	**98,744**	**59**

6.4 Economic Assessment of 300-kW Wind Turbine Project

6.4.1 Estimated Payback Period (PP) of Turbine Investment

Based on an annual energy output of 288,025 kWh units of electrical energy produced by the turbine in 2014, the following Payback Period (PP) is estimated. From previous electricity bill analysis, the business uses 72% of electrical energy during day hours and 28% during night hours. Day units cost €0.1925 per kWh and night units cost €0.0901 per kWh. Annual day monetary savings due to the turbine investment is €39,920 [(0.72 × 288,025) × €0.1925] and annual night savings is €7266 [(0.28 × 288,025) × €0.0901]. The total annual savings are €47,186. The turbine annual service and maintenance are €4270. As the initial capital cost is €280,000, the estimated Payback Period is **6.5** years [€280,000 ÷ (€47,186−€4270)].

6.4.2 Electricity Utility Energy Costs for SME

The data shown in Table 6.8 compares the SME monthly cost of electrical energy for the years 2013 and 2014. On analysis of the monthly comparisons, there appears little evidence to suggest a 50% saving as promised by the company selling the wind turbine. These figures indicate that there are little if indeed any monetary savings for the business since the introduction of this wind energy project, so this particular wind turbine installation is not reducing the electricity costs for its investor. The monetary values shown in Table 6.8 are the total amount payable by the customer. To differentiate between different types of costs on each bill, the November 2014 figure is used as an example, where each element is individually analysed.

- Energy (kWh) costs €14,088 (72%),
- Capacity charge (kVA) costs €2988 (15%),
- Public Service Obligation levy costs €1940 (10%),
- Energy management fee costs €571 (3%) and
- Total cost €19,587 (100%).

Note that there are no wattless unit (kVArh) penalties associated with this bill. The cost of the increased Maximum Import Capacity value from 320-kVA to 600-kVA

Table 6.8 SME monthly utility electric energy costs for comparison pre-connection (2013) and post-connection (2014)		2013	2014
	January	€15,423	€19,514
	February	€14,028	€17,275
	March	€13,537	€17,068
	April	€14,397	€17,609
	May	€15,804	€19,913
	June	€15,154	€17,627
	July	€20,941	€19,961
	August	€20,089	€18,372
	September	€18,227	€20,550
	October	€19,258	€19,119
	November	€20,974	€19,587
	December	€18,478	€18,902

was offset by the omission of the 'Excess Import Capacity' penalty charges before the kVA increase.

6.4.2.1 Energy Import Values with Turbine OFF/ON

As stated in Sect. 6.2.1, an opportunity arose to compare the energy imported to the factory while the turbine was switched OFF with the energy imported when it was switched ON, for the same period under similar working conditions. Those dates are

- Friday 30 May 8 a.m. to Tuesday 10 June 8 a.m. (turbine OFF) and
- Friday 13 June 8 a.m. to Tuesday 24 June 8 a.m. (turbine ON).

The switch-off was to facilitate necessary testing and maintenance. Table 6.9 reports on those imported energy values.

The values in Table 6.9 indicate an increase in all three imported units when the turbine was switched ON. Note that there is a Bank Holiday Monday on 2 June 2014 which is not a typical working day for this business. The actual recorded energy values for 2 June 2014 are 1855 Day kWh units and 1074 Night kWh units. Previous energy bills suggest an average Monday energy usage of 2791 Day kWh's, 821 Night kWh's and 2230 kVArh Wattless units so a slight adjustment must be made to these values. These average Monday working day calculations are obtained from 31 March 2014, 7 April 2014, 14 April 2014, 21 April 2014 and 28 April 2014 stored values. The workload in the factory is constant and very predictable over any seven-day period. Tables A6.1 and A6.2 (Appendices for Chapter 6) indicate the average daily wind speeds for the significant dates with the twelve-day average on the bottom of each column.

Weather data was obtained from the Irish Meteorological Service (Met Eireann 2015) for average monthly wind speeds for each month in 2014 and for May and June in 2013 to check if there were any unusual weather events or patterns to emerge during the test period examined. Analysis of the acquired data indicated that on the dates in question, the weather was typical of other similar dates so no external influences can be attributed to these unexpected results. Table A6.3 (Appendices for Chapter 6) identifies the wind speed values for this period.

The digital output indicator in the base of the turbine tower (Fig. A6.3) also indicated a significant number of instances over a protracted period where the turbine went into stop mode due to a *current deviation* error. This error initiated a dispute between the electric utility supply authorities and the turbine supplier as to the cause of the error. The electrical supply authority recorded the current value in Amps at the supply intake with a high-resolution meter that read values every milli-second. This

Table 6.9 Comparison of imported units with 300-kW wind turbine OFF/ON

	Wind turbine	Day units	Night units	Wattless units
30 May 2014–10 June 2014	OFF	28,572	13,716	11,892
13 June 2014–24 June 2014	ON	34,452	14,736	16,872

test was identified between 30 and 40 spikes over 1000 Amps, some over 1700 Amps over a 24-h time period. A possible cause of such spikes is the switching of large inductive loads. The turbine supplier also recorded the power, voltage and current at the supply intake between the periods 4 April to 11 April 2014. They claimed that even though the factory electrical load was steady between 80-kW and 100-kW, there were 100 V dips recorded during this period, and the turbine supplier thus laid the blame on the supply authority for these voltage dips. The final cause of the error was not definitely resolved, but the electric utility supply authority replaced the 400-kVA supply transformer with a 630-kVA supply transformer and the instance of the error occurring appeared to reduce following this action.

6.5 Discussion

6.5.1 GHG Emission Reduction Due to 300-KW Wind Turbine

One of the purposes of the wind turbine was to reduce the GHG emissions due to the importing of kWh units of energy produced mainly by fossil-fuel-driven generators. Table 6.3 indicates that this has not had the desired effect on GHG emissions. Note that 1 kWh of electrical energy emits 456.6 grammes of CO_2 (SEAI 2015, p. 82) into the atmosphere (Table A4.1, 2014 value). The SME has seen a constant rise in CO_2 emissions over the past five years, and the installation of the wind turbine did not thwart this trend. In 2016, there were 750.8 Tonnes of CO_2 emitted into the atmosphere due to operations requiring imported electrical energy from the National Grid. As can be seen from Table 6.3 (and Table 6.1), the number of kWh imported units continues to rise. The kWh/Tonne benchmark values in Table 6.1 are getting worse because in 2015, it required 20.90 kWh units of imported energy to produce one Tonne of product while in 2014 it required 19.04 kWh units of imported energy to produce the same output (one Tonne). This deterioration in the energy benchmarks should have raised a 'red flag' for the SME and for the government body charged with looking after the sustainability reporting mechanism, but it appeared not to do so. It is clear from the data presented in Table 6.1 that the DFIG turbine, in this case, is not an economically viable alternative to fossil-fuel-driven electrical generators as the number of kWh units imported from the, mostly fossil-fuel-driven, National Grid continues to rise unabated.

6.5.1.1 Highly Dispersant, Constantly Ramping Turbine Power Output Values

The digital output reading in the turbine tower appears to overstate the number of kWh energy units produced by the turbine by a significant amount, approximately

41%, compared to the analogue electromechanical rotating disc reading. The fast-changing, continually ramping power output signal is sampled at regular intervals to digitally calculate the energy, and it appears that it is not calculating this value accurately. A look at the energy benchmarks in Table 6.1 would seem to confirm the fact that the SME is not saving 460,842 kWh units of electrical energy annually due to the wind turbine contribution as suggested by the digital energy indicator in the base of the turbine tower. The rotating disc electromechanical kWh energy meter is designed in such a way that a force (driving torque) is exerted on the disc, the speed of which is proportional to the value of voltage, current and power factor in the circuit. It appears that the promised/expected/perceived reduction in CO_2 emissions due to the installation of this wind turbine is overstated when compared with the actual reduction. There is a 41% difference between the digital meter reading and the rotating disc meter reading in the number of kWh units of energy. This inaccuracy bears similarities with the VW scandal in Germany where the CO_2 emissions were not as stated by the company. On a micro level, the SME does not see a significant reduction in their electricity bill or the number of imported kWh energy units as a result of the wind turbine installation on-site which led to the idea of putting the rotating disc as shown in series with the digital meter. On a macro level, there is an increase in the amount of GHG emissions worldwide despite the promotion of wind turbines as alternative energy sources. Questions must be asked about the reliability of sustainability reporting frameworks and governance regarding the number of kWh energy units produced and measured by digital meters. The constantly ramping wind turbine output signal (Figs. 6.3, 6.4, 6.5 and 6.6 and Table 6.4) may also be contributing to overall disappointing results for the turbine investors as the ramping of an electrical generator can lead to undesirable effects with a lessening in expected benefits (Cullen 2013; Kaffine et al. 2013).

6.5.1.2 Statistics Associated with the Output from the Wind Turbine

Regulation (EC) No. 1099/2008 of the European Parliament is an essential tool for evaluating energy statistics within each member state. In this framework, Article 6 gives guidelines on quality assessment and reports and states explicitly that member states shall ensure the quality of the data transmitted. In this case study, the quality of the data transmitted leaves a lot to be desired. The governmental body tasked with this quality assessment needs to do more to improve these very valuable statistics. The EU Directive 2009/28/EU (Article 14) states that member states must make available information on the net benefits of generating electricity from renewable sources. This information is to be provided by either the supplier of the wind turbine or by the competent national authorities. The author of this book suggests a much higher input from governmental bodies as the equipment supplier, in this case, overstated the benefits when selling the wind turbine. This overstating may be down to several factors, among which may be (i) limited previous empirical results, (ii) lack of specialist knowledge in the area and (iii) desire to make a profit by supplying the turbine. The national authorities must ensure that academia is contributing to the raising of corporate social responsibility awareness by training personnel in this sphere. This lack of regulatory guidance is stated by Androde, and Puppim de Oliveira (2015) is also reported as one of the reasons for the failure to get GHG emissions down. Questions must be asked as to the effectiveness of the United Nations in governing and policing the global energy strategies among member nations. Perhaps it is time for a new global governance body in the area of GHG emissions, an agency that has influential and robust power to ensure that policies are being implemented in a meaningful and effective way.

6.5.2 Electricity Bill Cost Reduction with 300-kW Wind Turbine

Based on the findings of this study, the expected financial benefits of installing an embedded 300-kW wind turbine generator did not materialise for this business. The business took its CSR issues seriously by deciding on this strategic investment, but the promises and assurances from the company supplying the turbine did not live up to the expectations of this SME. From an ethical perspective, the failure of the turbine vendor to seek out the truth regarding the number of units that the turbine would produce on an annual basis cost this company significantly. The net cost of the project was €280,000, and this investment was made on the premise that there would be definitive cost savings for the company as well as fulfilling the desire to adopt green energy strategies within a social responsibility for the corporation. The comparison of the monthly electrical energy costs for 2013–2014 in Table 6.8 shows that there is no reduction in the company's electricity bill. Table 6.2 data indicates a production increase of 10% between 2013 and 2014 with a corresponding electrical energy increase of 12%. There is certainly no evidence to suggest that the company is saving 50% on their bills as promised by the wind turbine seller. Regarding this lack of monetary savings, this author was concerned that in this individual project the external energy management company failed to highlight the lack of savings, although they continue to be employed by the SME on an ongoing basis to supply the electricity and manage the energy usage within the business, charging 0.4 cents per kWh (approximately €500 per month) for their services.

One of the significant unexpected findings of this piece of research is the fact that the business imported more kWh units of electrical energy when the turbine was switched ON compared to the units imported for precisely the same days and time frame when the turbine is switched OFF (Table 6.9). Even allowing for a small correction factor in these values because of the Bank Holiday Monday, 2 June 2014, the results are contrary to what we expect them to be. The wind speed values stated in Tables A6.1 and A6.2 show that the period between 13 June 2014 and 24 June 2014 when the turbine was switched ON was not particularly windy, but this does not account for the unexpected results shown. The fact that the turbine had stopped on several occasions on a 'current deviation' error as discussed in Sect. 6.4.2.1 meant that on start-up, the machine acts as a motor which may account for some of the extra energy imported from the National Grid during this start-up period. The source of the 'current deviation' error is not definitive, and a recording and analysis of the power output of the turbine may yield some fresh insight into this problem.

The connection of the inductive wind turbine in parallel with the electricity supply appeared to damage the overall power factor of the installation. This deterioration is confirmed by the results shown in Table 6.6. The deterioration necessitated the purchasing of a Power Factor Correction (PFC) unit for €4000 by the business owners. In our study, the low power factor required the increase of the MIC value, which also has a cost associated with it. Data in Table 6.9 indicates a substantial increase in kVArh (Wattless) units of 42% when the turbine is embedded with the National Grid, i.e. switched ON, compared with the wattles units when the turbine is not connected, i.e. locked OFF. At the request of the external energy management company, the factory MIC was increased from 320-kVA to 600-kVA, and the size of the supply transformer was increased from 400-kVA to 630-kVA. This result is not the expected outcome of installing a second embedded source of power/energy, as the expectation is that less imported power/energy is required to power the business premises.

As the variable-output wind turbines require gas plants as 'spinning reserve' (SEAI Report 'Quantifying Irelands Fuel and CO_2 Emissions Savings from Renewable Electricity in 2012'), the advantages of having wind turbines is very much questionable particularly as the evidence in this case study suggests there is limited if any decrease in kWh units imported to the consumer in this case. The constantly ramping wind turbine output signal (Figs. 6.3, 6.4, 6.5, and 6.6 and Table 6.4) may be contributing to these disappointing results as ramping of the electrical generator can lead to undesirable effects with a reduction in expected benefits (Cullen 2013; Kaffine et al. 2013). The benchmarks used in the Origin Green sustainability plan in Table 6.1 showed only a very slight improvement in the kWh/tonne of product values, which is very disappointing. For example, the calculated values are 19.87 kWh/tonne for 2014 (following the installation of the wind turbine) and 19.83 kWh/tonne for 2013 (before the wind turbine was installed). This finding would appear to agree with Milne and Gray (2013) who claim that some sustainability reporting businesses do not change their processes or methods or become more sustainable and are content to carry on with the status quo. Sustainability reporting is not an end in itself, but an indication of progress or otherwise in the sustainability efforts of the business. Sustainability reporting benchmarks should be an alarm signal for managers to investigate the lack of progress in certain process areas.

6.6 Conclusions

6.6.1 Effectiveness of Government Energy Policy

Government energy policy of promoting wind turbines as a '*green*' alternative generator of electrical energy failed in this case to achieve its goal, namely to reduce GHG emissions. The SME did not significantly reduce the number of imported kWh electrical energy units purchased via the National Grid despite the claim by the turbine developer that the turbine would produce 600,000 kWh units of electrical energy per year. Subsequently, the initiative did not reduce the CO_2 (GHG) emissions as the Irish government strives to meet EU-binding emission targets. The measured power output data from the wind turbine presented in Figs. 6.3, 6.4, 6.5, and 6.6 demonstrates a constantly ramping power output signal with a high coefficient of variation values (Table 6.4). The poor quality of the wind turbine power, particularly the short-term dispersion effect, is creating problems for the parallel-connected fossil-fuel generators. The fossil-fuel generators are unable to respond fast enough to compensate for the wind turbine variations. The impact of this is that there are no savings (reductions) associated with the number of kWh electrical energy units imported from the National Grid. The spinning reserve embedded conventional back-up generators are unable to compensate for the fluctuating, ramping output of the wind turbines. Cullen (2013) suggests that ramping of traditional (fossil-fuel) electricity generators may increase their emission rates (Cullen 2013). The findings of this current study agree with the results by Kaffine et al. (2013) that emissions savings may be eroded entirely in some cases due to cycling-related emissions. Worryingly, the research highlighted a flaw in the sustainability reporting mechanism and the governance associated with such a process. The benefits of the turbine were greatly overstated, and the erroneous readings went undetected despite a plethora of state bodies having access to the data.

6.6.2 Stakeholder Outcomes in 300-kW Wind Turbine Investment Decision

The only stakeholder who appeared to have benefitted from the project is the turbine supplier who got paid in full from the SME for providing the turbine. However, research findings such as this piece should inform academics and practitioners in the alternative energy debate and move it on from the 'group-think' mentality that appears to exist currently in the Irish narrative. For wind turbine suppliers to develop their businesses sustainably, they must provide an improved product or service than was afforded to the SME in this study. If turbine suppliers are 'unsure of outcomes' from the investment, then from an ethical perspective there may be an onus and moral obligation on their part to inform the investor of such uncertainties (Frederick 1999). This study provides managers and business owners with informed and independent data by which to question and critique the benefits of such future investments and make better, and more profitable decisions.

Instead of being a profitable project for this SME, the connection of the embedded wind turbine made things worse for the company. Indeed, it appears that the wind turbine affected a necessary increased MIC and also a deterioration of the power factor (Table 6.6 and Fig. A6.2) with no apparent benefits in terms of reducing the cost of their electric energy bills (Table 6.8). Several meetings with the renewable energy turbine provider and adoption of numerous changes to turbine settings have to date failed to yield benefits, and the SME has resigned itself to a 'wait and see' attitude. One of the four critical inputs identified in Sect. 3.5.3 for businesses to develop sustainably is that CSR initiatives must be profitable. This €280,000 investment was not profitable and therefore, is not contributing to sustainable business development in this case. An investment that is not profitable and produces no assistance at protecting the environment does not fulfil the social dimension of sustainability (Kuhnen and Hahn 2019). Interestingly in this research, the external energy management company employed by the SME to supply the electricity and manage the energy usage failed to 'flag' any noticeable link between the number of kWh energy units that imported pre-turbine and post-turbine installation. This kWh value would have been expected to decrease post-turbine installation but failed to do so. Concerning the deterioration of the power factor, the increase of the MIC value from 320-kVA to 600-kVA may have only served to mask the problem, without actually identifying any underlying issues that may have contributed to the problem initially. The deterioration in the overall factory premises power factor after the wind turbine was connected (approximately 0.77 lagging) compared to before the turbine was connected (approximately 0.92 lagging) was not identified by any of the stakeholders involved. Note that the network operator ESB Networks states that the power factor of generators exporting energy to the grid should be in the range of 0.92 (ESB Networks 2017b). The robustness of the energy management function of this external energy management company is, therefore, questionable. Contributing factors to this apparent failure may be naivety on the part of the SME itself, a lack of knowledge/interest on the part of the external energy management company or most likely a skills gap in this emerging industry.

On the positive side for the company who made this investment, the aesthetic nature of the on-site wind turbine has been used in some marketing material as the company is perceived to be engaging with their corporate social responsibilities strategically, albeit in a purely aesthetic manner. The researcher's interaction with the company employees on the company premises allowed the researcher to perceive a feel-good factor due to the on-site wind turbine rotating in the wind.

6.6.3 Alternatives to Supply-Side Management

There is some research to show that a Building Management System (BMS) investment with smart metering has the potential to decrease the power demand (Ansani

2015) using an efficient way to schedule the use of power. It is recommended that further research is carried out to compare the Supply-Side Management philosophy of wind turbine installation with a Demand-Side philosophy '*smart*' installation on this same site. One of the problems of utilising wind turbines as a Supply-Side Management strategic tool (as described in this research study) is that it appears to be definitively difficult in capturing energy from wind in a steady, stable fashion. If the turbine power output is proportional to the wind speed (which is generally the expected case), then the localised wind speed is varying significantly with perhaps no small amount of turbulence. This finding is very significant as it appears that the quality of the power being produced by the turbines in this study may be reduced and falls short of perceived expectations and values. For example, ESB Networks (2017a) states that deviations in the ramp rate of electricity-generating turbines should not exceed 3% of registered capacity, a value far exceeded in this case study.

References

Agudo-Valiente, J. M., Garces-Ayerbe, C., & Salvador-Figueras, M. (2015, January/February). Corporate social performance and stakeholder dialogue management. *Corporate Social Responsibility and Environmental Management, 22*(1), 13–31.

Andrade, J. C. S., & Puppim de Oliveira, J. A. (2015, August). The role of the private sector in global climate and energy governance. *Journal of Business Ethics, 130*(2), 375–387.

Ansani, G. A. (2015). Role of smart meter in demand side management for future smart grids. *International Journal of Technology and Research, 3*(1), 7–14.

Aspling, A. (2013). Business, management education, and leadership for the common good. In D. L. Everett (Ed.), *Shaping the future of business education: Relevance, rigor and life preparation* (pp. 40–58). New York: Palgrave Macmillan.

Boutsika, T., & Santoso, S. (2013). *Quantifying the effect of wind turbine size and technology on wind power variability*. Power and Energy Society General Meeting, IEEE 2013 (pp. 1–5). https://doi.org/10.1109/PESMG.2013.6672587.

Catalin, C., & Nicoleta, R. (2011). International biomass trade and sustainable development: An overview. *Annals of the University of Oradea, Economic Science Series, 20*(2), 47–54.

Cullen, J. (2013). Measuring the environmental benefits of wind-generated electricity. *American Economic Journal: Economic Policy, 5*(4), 107–133. https://doi.org/10.1257/pol.5.4.107.

DCENR. (2014). Department of Communications, Energy and Natural Resources. Available at http://www.dcenr.gov.ie. Accessed 12 March 2015.

Doha. (2012). *Doha amendment to the Kyoto protocol*. Available at http://unfccc.int/kyoto_protocol/doha_amendment/items/7362.php. Accessed on 20 October 2014.

Elkington, J. (1997). *Cannibals with forks: The triple bottom line of 21st century business*. Gabriola Island, BC: Capstone.

ESB Networks. (2017a). Distribution Code, Version 5.0, April 2016, Section DCC11.3.4.1. Available at http://www.esbnetworks.ie. Accessed on 28 November 2017.

ESB Networks. (2017b). Distribution Code, Version 5.0, April 2016, Section DCC11.4.3. Available at http://www.esbnetworks.ie. Accessed on 28 November 2017.

Frederick, R. E. (1999). *A companion to business ethics*. Oxford, UK: Blackwell Publishers Ltd.

Gray, D. E. (2009). *Doing research in the real world* (2nd ed.). London: Sage.

Hansen, E. G., Zvezdov, D., Harms, D., & Lenssen, G. (2014). Advancing corporate sustainability, CSR, and business ethics. *Business & Professional Ethics Journal, 33*(4), 287–296.

Holme, L., & Watts, P. (2000). *Corporate social responsibility: Making Good Business Sense*. Geneva: World Business Council for Business Development.

Kaffine, D. T., McBee, B. J., & Lieskovsky, J. (2013). Emissions savings from wind power generation in Texas. *The Energy Journal, 34*(1), 155–175. https://doi.org/10.5547/01956574.34.1.7.

Kealy, T. (2014, October). Sustainable business development: An Irish Perspective. *International Journal of Humanities and Social Science, 4*(12), 166–179.

Kivunja, C., & Kuyini, A. B. (2017). Understanding and applying research paradigms in educational contexts. *International Journal of Higher Education, 6*(5). https://doi.org/10.5430/ijhe.v6n5p26.

Krisnawati, A., Yudoko, G., & RosBangun, Y. (2014). Development path of corporate social responsibility theories. *World Applied Sciences Journal*, 110–120. IDOSI Publications.

Kuhnen, M., & Hahn, R. (2019). From SLCA to positive sustainability performance measurement: A two-tier Delphi study. *Journal of Industrial Ecology, 23*(3), 615–634. https://doi.org/10.1111/jiec.12762.

Kyoto. (1997). *Kyoto Protocol, United Nations framework convention on climate change*. Available at http://unfccc.int/kyoto_protocol/items/2830.php. Accessed on 20 October 2014.

Lin, W., Wen, J., Chang, S., & Lee, W.-J. (2012, May/June). An investigation on the active-power variations of wind farms. *IEEE Transactions on Industry Applications, 48*(3), 1087–1094. https://doi.org/10.1109/TIA.2012.2190817.

Met Eireann. (2015). *The Irish meteorological service*. Available at http://www.met.ie/. Accessed on 7 March 2015.

Milne, M. J., & Gray, R. (2013). W(h)ither ecology? The triple bottom line, the global reporting initiative, and corporate sustainability reporting. *The Journal of Business Ethics, 118,* 13–29.

National Standards Authority of Ireland, I.S. EN 61400-1: 2005. *Wind turbines—Part 1: Design, requirements*.

Origin Green. (2017). Available at http://www.origingreen.ie/about/origin-green-promise/. Accessed on 4 November 2017.

RIO. (2012). United Nations Conference on Sustainable Development. Available at http://www.uncsd2012.org. Accessed 12 March 2015.

Roy, S. (2013, December). Power output by active pitch-regulated wind turbine in presence of short duration wind variations. *IEEE Transactions on Energy Conversion, 28*(4).

Shafiee, S., & Topal, E. (2009). When will fossil fuel reserves be diminished? *Energy Policy, 37*(1), 181–189.

Sustainable Energy Authority of Ireland. (2013). *Renewable energy policy*. Available at http://www.seai.ie/Renewables/Renewable_Energy_Policy/. Accessed on 12 March 2015.

Sustainable Energy Authority of Ireland. (2015). *Energy in Ireland 1990–2014, 2015 report*. Available at http://www.seai.ie. Accessed on 8 December 2015.

Teoh, W. L., Khoo, M. B. C., Castagliola, P., Yeong, W. C., & Teh, S. Y. (2017, February). Run-sum control charts for monitoring the coefficient of variation. *European Journal of Operational Research, 257*(1), 144–158. https://doi.org/10.1016/j.ejor.2016.08.067.

Thuita, F. M., Pelto, G. H., Musinguzi, E., & Armar-Klemesu, M. (2019). Is there a "complimentary feeding cultural core" in rural Kenya? Results from ethnographic research in five counties. *Maternal and Child Nutrition, 15*(1), 1–8. https://doi.org/10.1111/mcn.12671.

Wan, Y. H. (2004, September). *Wind power plant behaviors: An analyses of long-term wind power data* (Technical Report, National Renewable Energy Laboratory). Colorado, USA. https://doi.org/10.2172/15009608.

Yin, R. K. (2003). *Case study methods: Design and methods* (3rd ed). Thousand Oaks, CA: Sage.

Chapter 7
Key Enablers/Inhibitors in the Corporate Social Responsibility—Business Strategy Integration Space

Abstract Despite the potentially positive image and reputation implications of businesses implementing strategies in Corporate Social Responsibility (CSR), there appears to be a dearth of companies able to play a leading role in advancing CSR activities to upper levels within the cognisance of the organisation. The vast majority of businesses appear to be merely complying with national regulations in their business sustainability efforts. This study aims to investigate the key enablers and inhibitors in assisting businesses integrate their CSR with their business strategy. A 17-question online survey was administered to many national and global companies in a range of industries. The resulting qualitative and quantitative data from 86 respondents was analysed and is presented in tabular, graphical, and text form. Quantitative responses were presented in the form of descriptive statistics, while qualitative data was analysed and presented thematically. The study found that the two main enablers/inhibitors to businesses integrating CSR with their corporate strategy were (i) improved measuring/reporting techniques for evaluating CSR outcomes (ii) informed education/knowledge in all aspects of CSR. Both academia and the business world can contribute and assist in confronting the two main issues and guide businesses as they attempt to integrate CSR more closely with the business model of their organisation. The study should help to encourage business and engineering schools in academic institutions in designing educational programmes to include CSR learning outcomes to their programmes.

Keywords Corporate Social Responsibility · Strategy · Sustainability barriers · Enablers

7.1 Introduction

Evaluation of the role of business in society has evolved over the past couple of decades. Accountability and transparency in all aspects of business activity have augmented terms such as Corporate Social Responsibility (CSR), Corporate Responsibility (CR) and Corporate Citizenship (CC) in the business consciousness (Aspling 2013). This new reality has forced businesses to reconsider their role and duties in society and in the broader economic community. Catalin and Nicoleta (2011) claim

© Springer Nature Switzerland AG 2020 145
T. Kealy, *Evaluating Sustainable Development and Corporate Social Responsibility Projects*, https://doi.org/10.1007/978-3-030-38673-3_7

that the overall concept of sustainable development, that includes CSR, CR, and CC, emerged as an alternative philosophy to the neoliberal school of thought which has dominated economics for the past three decades, which argues for markets to be free with little or no intervention by government. This economic agenda was promoted by Ronald Reagan and Margaret Thatcher in the 1980s. This neoliberal economic theory is growth-driven, and while it is probably a reasonably robust business growth model, continuous growth has put pressure on our vast but finite natural resources causing environmental degradation which, according to the Catalin and Nicoleta (2011), can threaten our wealth and even our existence. In this study, key issues are identified that, with the help of academic institutions and the business community, empowers a business to embrace the new sustainability reality.

Hack et al. (2014) state that there are many definitions of CSR, while Holme and Watts (2000) declare that 'Corporate Social Responsibility is the continuing commitment by businesses to behave ethically and contribute to economic development while improving the quality of life of the workforce and their families as well as of the local community and society at large'. On a similar theme, Sustainable Development (SD) is often defined as *'development that meets the needs of the present without compromising the ability of future generations to meet their own needs'*. This definition emerged from the Bruntland Commission, which was set up by the United Nations (Bruntland Commission Report 1987) in 1983. Their mission was to direct sustainable development on a global level. The process of globalisation opens up new unprecedented opportunities of large-scale redistribution of wealth, and in such actions, humanity itself becomes increasingly interconnected. The UN subsequently initiated eight international development goals in the year 2000 entitled the UN Millennium Development Goals, and the objective was to reach these goals before the year 2015. Along with the eight goals, there were 21 targets and a series of measurable indicators to assess if the sustainability targets were being met or not. Questions remain as to whether the UN Millennium Development Goals have been reached (Sandbu 2015). A follow-on framework was put in place with the setting up of the Sustainable Development Goals (SDG) to cover the 15 years between 2015 and 2030. The new proposal contained 17 goals and 169 targets. Implementation of the structures to ensure that the goals are achieved was discussed at the 2015 United Nations Climate Change Conference (COP21) held in Paris between 30 November and December 11 2015. There is significant interest in the environmental dimension of the SDG framework and in particular, the impact that business activity is having on the phenomenon of climate change (Halati and He 2018).

Currently, there is a proliferation of opinions, rhetoric and studies from a plethora of stakeholders regarding the topic of climate change. These are emanating from many sources even from outside of business and politics. Recently opinions and comments have emerged from leaders of some of the largest religions in the world (Pope Francis 2015; Ali 2016). The 'encyclical' *Laudato Si* (Pope Francis 2015) was published in 2015 and swiftly followed by the publication in Istanbul of "Islamic Declaration on Global Climate Change" (Geographical 2015), from the Islamic leaders of 20 countries worldwide. These faith-based instructions to millions of followers worldwide appear to have evolved in tandem with societal claims that there was pertinent need to examine human and social effects on environmental degradation is contributing to the climate change phenomena. Pope Francis (2015) and Ali (2016) claims that religion and science, with their distinctive approaches to a philosophical understanding of reality and human existence, can enter into an intense dialogue to address climate change issues. Commenting on religious/spiritual leaders entering this climate change debate (Li et al. 2016) declared that the issue of climate change is such a human imperative to currently address, that there should be room for different diverse pathways by which individuals with different cultural values and political and religious views can come to a shared belief, dialogue and action regarding the need for action on climate change. They claim that it is clear that positive influence and action on climate change is now an imperative for leaders of all facets of society that inhabit the globe (Li et al. 2016).

Businesses are significant stakeholders and have a responsibility and opportunity to make responsible decisions and therefore play their part in protecting the natural (and social) environment. How businesses source and use their electrical energy is one facet by which they can contribute to the protection of the environment (Schwerhoff and Sy 2017). The area of CSR is generally considered to belong to the academic discipline of Management. The upper (senior) level in the business management hierarchy is where strategic decisions are made (Smith 2014). This study seeks to determine if businesses are actually embracing the economic, social, and environmental dimensions of business management decisions and if not, then the enablers and barriers that need to be addressed to enhance responsible corporate development are identified. It is envisioned that the identified enablers and barriers assist businesses to progress from the lower level of CSR engagement (doing the minimum to comply with regulations) through the middle level (report voluntarily and strive for sustainability rankings) and on to the upper level (engage in responsible actions for the common good as explicit part of their system) (Aspling 2013). It is at the upper level that companies effectively integrate sustainability strategies into their business model, i.e. its value creation process (Bini et al. 2018). A framework for assessing CSR engagement strategies is presented by Tang et al. (2012) which provides a general guideline for practitioners in engaging CSR activities.

7.2 Literature Review

7.2.1 Overview/Definitions

The Bruntland Commission Report (1987) for the World Commission for Environment and Development defined sustainable development as '*development that meets the needs of the present without compromising the ability of future generations to meet their own needs*'. Increasingly from a European perspective, the concept of sustainability is grouped under the 'umbrella' term of CSR (Jones Christensen et al.

2007). In the past decade, the concept of CSR is commonly used for what in the past may have been a collection of terms such as corporate accountability, corporate citizenship, triple bottom line, social performance, etc. (Silberhorn and Warren 2007). Commenting on overall perceptions of the concept of CSR and sustainability, Gobbels (2002) declares it a somewhat 'fuzzy notion' within the business community. He maintains that many attempts to give CSR a succinct universal definition have been equalled with diverse and varying degrees of perception of the concept of CSR. However, theories on what constitutes CSR continue to evolve within various conceptual frameworks. More recently these initial concepts of social effects and ethical behaviour of an organisation under the CSR realm have evolved, to include in-depth analysis of the performance of the organisation also, and more specifically evolved to include more detailed analysis of CSR's effect on profit (Lee 2008). It appears that increasingly CSR is now being presented within the concrete business strategy paradigm in response to stakeholder pressure and performance considerations (Silberhorn and Warren 2007). Chen et al. (2018) claim that CSR behaviour or indeed irresponsible corporate behaviour can benefit or indeed imperil corporate financial performance under certain conditions, but they claim that CSR and financial performance are inextricably linked.

In a much-quoted and seminal paper, Elkington (1997) identified three component areas of sustainability (Social, Economic, and Environmental, Fig. 7.1) namely the Triple-Bottom-Line framework (TBL) (Elkington 1997). However commenting on the TBL concept, Norman and MacDonald (2004) noted, '*there is no careful definition of the concept*', '*only vague claims about the aims*' (Norman and MacDonald 2004). This triple bottom line of social, economic and environmental dimensions was highlighted and adopted by Jones et al. (2011) in preliminary investigations

Fig. 7.1 Triple-Bottom-Line framework (Elkington 1997)

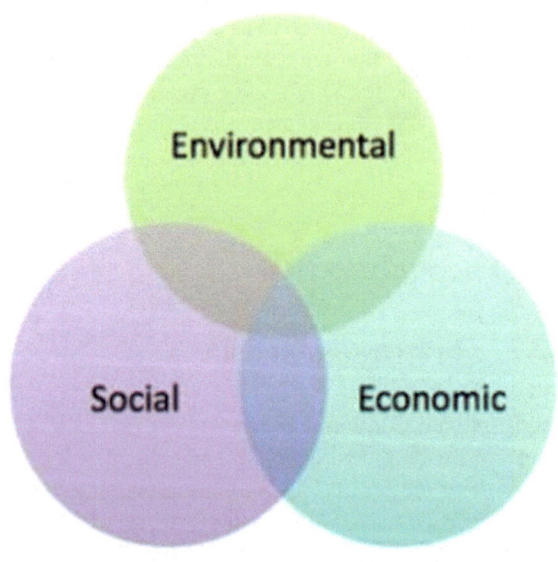

into the sustainability agendas of the world's top ten retailers, through an examination of Marks and Spencer's (M&S) sustainability strategy. These authors argue that the triple bottom components 'provide a robust empirical frame of reference' for sustainability strategy development.

However, increasingly commentary and robust criticism have emerged regarding shortcomings in the TBL framework and its conservative failure to evolve from often competing imperatives of profitability, environmental protection and social justice (Sridhar and Jones 2013). Sridhar and Jones (2013) argue that this TBL framework needs to adapt and change within the business economic paradigm to better integrate the almost separate disciplines of 'social and natural sciences' with the economic imperatives of the business organisation. Milne and Gray (2013) highlighted the dangers of confusing sustainability with the TBL framework while Rambaud and Richard (2015) proposed shifting away from the TBL towards what they argue should be a more genuine integrated reporting system, because they claim the TBL model based on eco-efficiency is not a guarantee of progress for environmental issues. Other authors have highlighted the justifiably questionable reductionist assumption of the whole environmental sustainability issue in TBL theory, i.e. the assumption that if each singular discrete firm is deemed to be 'sustainable', then overall planetary sustainability is achieved (Ozgur and Hernke 2017; Savitz 2012).

From the definitions of CSR and sustainable development, it is clear that many dimensions must be considered in business management decisions. Of the three dimensions described by Elkington (1997) (Fig. 7.1) financial reporting is the most established entity however more recently the environmental and social aspects are becoming more prominent (Cubas-Diaz and Martinez Sedano 2018). As a result of the changing global landscape, several standards (e.g. the UN Global Compact, the OECD Guidelines for Multinational Enterprises, the Global Reporting Initiative, Origin Green) have emerged to help corporations implement, manage and report their CSR activities (Vigneau et al. 2015). Issues surrounding the completion of such sustainability reports are analysed in this research.

Research by Pless et al. (2012) and Waldman (2011) also implies that leadership is essential for advancing the challenges and demands associated with the integration of CSR activities within a firm. However, Pless et al. (2012) suggested that managers interpret and display leadership in many ways, and this may contribute to the lack of evidence that leadership have on firm-level and societal-level outcomes.

7.2.2 Measurement and Reporting

Sustainability reporting and the need for businesses to actively engage in the process have intensified over the past thirty years. This has in part been due to increased legislative regulations but has also been dictated by more intense stakeholder engagement and demands for increased transparency in CSR (Adams and Evans 2004; Taylor et al. 2018). A study on 250 firms listed in the Fortune Global list (Fortanier et al. 2011) found that the existence of global standards and guidelines on sustainable issues not only increases the overall level of CSR reporting but also assists in a more uniform harmonised approach to the worldwide reporting of CSR activities of firms from different countries. This, they claim negates the divergent issues from local CSR domestic legislative and societal demands and reduce what they term 'the country of origin' effect on CSR reporting. Their findings indicate that global voluntary guidelines may have an important role to play in CSR reporting in addition to also to local domestic legislation (Fortanier et al. 2011).

Some authors are increasingly asking 'does a robust reporting mechanism indicate accurate CSR activity measurement?'. There are claims that the past twenty years have seen a lack of uniformity and clarity in CSR theoretical measurement methods (Venturelli et al. 2017). Increasingly there are justifiable arguments being introduced regarding the integration of sustainability measurement into the strategic business model in order for measurement to be more robust and succinct (Engert et al. 2016) The concern appears to be that the plethora of current reporting CSR paradigms and unmonitored regulation are dictating a proliferation of green-washing activities (or blue washing from global UN perspective) and may not at all be equalled by genuine, accurate company CSR measurement data (de Vries et al. 2015). Cubas-Diaz and Martinez Sedano (2018) proposed the Relative Sustainable Performance Measure (RSPM) and the Measure of Commitment failure (MC) that permits sustainable decision-making using the TBL approach. The two measures take into account the environmental and social variables in addition to the economic variables. The authors claim that it enables companies to analyse their sustainability performance and adapt their business plans accordingly.

7.2.3 Global Sustainability Initiatives

In terms of global business sustainability initiatives, the United Nations (UN) launched the Global Compact initiative in 2000 to encourage businesses worldwide to implement sustainable and socially responsible policies and report on their implementation. The Global Compact (United Nations—Global Compact 2014) is based on ten guiding principles, covering the areas of human rights, labour, the environment, and anti-corruption. The two objectives of the UN Global Compact are to (i) mainstream the ten principles in business activities around the world, and (ii) catalyse actions in support of broader UN goals, such as the eight Millennium Development Goals, targeted to be achieved by 2015. There appears to be some criticism to the UN Global Compact initiative as it appears that the initiative has not had the desired effect on global businesses. Some of the criticism stems from the fact that the Global Compact is not a regulatory instrument and so has no enforcement provisions, thereby rendering it ineffective in holding businesses accountable. It appears to operate more like a discussion forum which can leave itself open to misuse by businesses who can promote themselves as sustainably-responsible businesses without actually making any significant change to their strategic efforts as claimed by Milne and Grey (2013).

Luthra et al. (2018) identified the key issues that influence sustainability initiatives to be implemented in global businesses. The three most prominent issues identified were (i) government support systems, (ii) knowledge and awareness of sustainability tools, and (iii) information systems network design. However, it is also claimed that one of the main challenges of the sustainability agenda is that business engagement with the process appears to be primarily rooted within the corporate reporting realm as opposed to concrete sustainable activity (Gray 2010; Milne and Gray 2013).

7.2.4 CSR/Sustainable Development Business Initiatives in Academia

CSR/Sustainable Development concepts have originated mainly from the management area (Carroll and Shabana 2010; Carroll 1979). The CSR theme has been explored from various management perspectives such as social obligations, marketing, stakeholder-relation, integrated strategy and leadership themes. Senior leaders operate at the strategic, upper, level of an organisation (Smith 2014). The difference between leaders and managers has been expressed by the Austrian-American economist Peter Drucker (1909–2005), who claims that senior personnel may be 'doing things right' (managers) rather than 'doing the right things' (leaders) concerning CSR/sustainable development practices within a firm. The characteristics and traits possessed by business leaders appear to be crucial to the development of sustainable business practices within that organisation (Pless et al. 2012). Increasingly however it is clear that investors are now taking into account variables other than the economic ones of profitability and risk, namely those issues pertaining to environmental and social variables (Cubas-Diaz and Martinez Sedano 2018). Waldman et al. (2006) consider how a CEO's leadership style impacts on the company's' CSR values and that the CEO's vision of CSR may infiltrate to lower-level managers in the organisation. Academia and Business Schools, in particular, have a major role to play in helping to foster and develop such merits in their future leaders. The UN appears to agree with this statement and 2004 the UN Global Compact along with the European Foundation for Management Development (European Foundation for Management Development 2014), formed the 'Globally Responsible Leadership Initiative' (GRLI). The GRLI (Globally Responsible Leadership Initiative 2014) seeks to influence and guide academia so that business schools' mission, strategy, and activities show evidence of its contribution to ethics and sustainability. It is a worldwide partnership of companies and business schools working together to develop the next generation of globally responsible leaders. It may be that it is not enough that business schools limit their curriculum to the traditional core business disciplines of Marketing, Accounting, Strategy, Finance, Management and Economics but imperative that they also include embedded modules on sustainable development/corporate social leadership (Seto-Pamies and Papaoikonomou 2016). A hypothesis of this research is that there is a need to include CSR/SD modules in the business school curriculum with a major focus on non-financial reporting (human and environmental aspects in Fig. 7.1). Studies by Jones Christenson et al. (2007) and Medina Rivilla and Medina Dominguez (2014) suggested that sustainable development requires a better and more fruitful interface between academia and the business world. They recommend that learning institutions focus on competencies such as leadership, planning, management, communication, motivation, methodology, resource optimisation, student collaboration, innovation, inquiry, reflection, and empathy and that lecturers must be prepared to assist in creating an environment conducive to entrepreneurship. Research by Smith and Ronnegard (2016) examines the Shareholder Primacy Norm (SPN) as a widely acknowledged impediment to CSR. Smith and Ronnegard (2016) suggest that SPN is the part of the manager's legal fiduciary duty that requires them to make decisions that further the interests of the shareholders which stems from a largely unquestioned adherence to shareholder theory in business schools. The efforts of the GRLI initiative are certainly not apparent judging by the comprehensive report in 2014 by the Boston Consultancy Group (BCG), MIT Sloan Management Review and the United Nations Global Compact entitled 'Joining Forces: Collaboration and Leadership for Sustainability'. Of the 3795 senior and middle managers from 113 countries interviewed, only 20% of respondents believe that their senior-level executive boards provide substantial insight on sustainability issues. The report goes on to say that one of the strongest barriers to greater executive-level engagement is a lack of sustainability expertise among board members. If this is the case, then perhaps a better and more fruitful interface between academia and the business world, as suggested by Medina Rivilla and Medina Dominguez (2014), would assist in developing the tools required by business leaders to address sustainability issues. A review research paper by Carollo and Guerci (2017) explores how CSR managers use rhetorical styles and continually shift chameleon-like between them, to maintain a level of perceived primacy in the value they add to an organisation. They claim than rather than acting as the change agents that their roles could allow, CSR managers tend to err on the side of corporate caution and preoccupy themselves with strategic marketing efforts to secure their continued survival in the corporate landscape (Carollo and Guerci 2017).

Of the three dimensions identified in Elkington's (1997) Triple Bottom Line theory (Fig. 7.1), the financial aspect is perhaps the most established dimension taught in academic institutions. The other two non-financial (environment and human) dimensions are probably less well established on the academic curriculum. As a result of the changing global landscape, many standards (e.g. the UN Global Compact, the OECD Guidelines for Multinational Enterprises, the Global Reporting Initiative, Origin Green) have emerged to help corporations implement, manage and report their full CSR activities (Vigneau et al. 2015). Issues surrounding the completion of such sustainability reports (Sartori et al. 2017) are analysed in this research.

7.3   Methodology

The study aimed to target a cross-section of businesses with an established attentiveness to CSR who reside on national (Irish) and international sustainability reporting databases. A survey research methodology was used in this study, where data was obtained using an on-line questionnaire. The survey consisted of seventeen question questionnaire tool. The questions were developed from pertinent issues within the literature. A pilot study was conducted initially to ensure the robustness of the tool and the questionnaire was distributed to two experts in the field before distribution. It was decided to utilise closed-type and open-type questions to collect quantitative and some qualitative data, respectively. This research approach was chosen because it is believed that it could obtain pertinent data from a large sample. It was anticipated that this mixed-method approach would identify reasons as to why businesses appear to be finding the process of integrating a robust CSR paradigm with the strategy of the organisation very challenging. Several studies have identified these challenges (Isil and Hernke 2017; Hussain et al. 2018; Chen et al. 2018; Cubas-Diaz and Martinez-Sedano 2018).

This on-line survey method minimised the complexities of administering a questionnaire across different geographical locations to businesses participating in this research. The 17-question survey was designed using simple and clear language to reduce the potential for ambiguity by the survey respondents. Quantitative responses were presented in the form of descriptive statistics, while qualitative data was analysed and presented thematically. Thematic analysis is a means of analysing qualitative data in a rigorous and methodical manner (Nowell et al. 2017). Thematic analysis was utilised by developing codes from the survey data (Braun and Clarke 2006). The coding process allows for the simplification and focuses on pertinent characteristics within the data. Subsequently, themes are developed from the codes. Classified themes are used to identify patterns that underlie the themes. The cross-tabulation method of data analysis was utilised to quantitatively analyse the relationships between the range of factors considered in this study.

7.3.1   Data Collection

During the initial contact with potential businesses, the appropriate liaising person was identified to whom the survey was sent by e-mail with a link to the questionnaire web-page. Survey respondents were assured that their answers would be confidential. Respondents who agreed to take part in the survey were notified in advance about the impending e-mail with a link to the survey. Because of the contact with subjects during the survey process, care was taken to minimise informal manipulation of the data, as stated by Yin (2008) and Zivkovic (2012). The questions are listed in Appendix 7A. The survey included questions of both the qualitative and quantitative type. In choosing respondents, a random sample of the business community was taken from both national (Irish) and international sustainability-reporting databases. Such companies would appear to have CSR practices embedded in their business strategy. All of the verified members of Bord Bia's (Ireland) Origin Green sustainability reporting framework were contacted to take part in the survey, $n = 104$. Of this total value, 35 businesses completed the survey on-line. After this, international companies were randomly chosen from the Global Reporting Initiative (GRI) sustainability database to participate in the research study, $n = 290$ firms. These were contacted by phone and e-mail to request their participation in the study. In total, there were 394 company contacted (national and international) with 86 completing the survey on-line. This gives an overall completion rate of 22%.

7.3.2   Data Analysis

The quantitative data is analysed and represented in tabular and graphical form. Analysis of the qualitative text data was conducted using the thematic approach (Braun and Clarke 2006). Analysing text presents a challenging task for qualitative researchers. It is claimed that a detailed description of how researchers identify themes from the qualitative data relies somewhat on intuition which is challenging to describe making theme identification an under-reported concept (Vaismoradi et al. 2016). However, validity and reliability can be applied to qualitative data analysis if step-by-step stages are rigorously followed (Braun and Clarke 2006). The qualitative thematic analysis of the open ended questions (qualitative text data) for this study included the following five stages; firstly, the data was organised into a word document format and stored on the researchers PC where the researcher read the transcripts in their entirety to get a sense of the main points being made before the second stage began; secondly, key words in context were identified to allow coding of the data (and assigning labels to each code) whereby the text was grouped into small categories (Vaismoradi et al. 2016) of information (the results of this stage can be viewed in Appendix 7); thirdly, broad units of information called themes(s) were identified based on a grouping of the codes from stage two that form a common idea; fourthly, the researcher made sense of the data by interpreting the codes and themes into an overarching meaning of the data (the interpretation is based on the researchers personal views *and* the research literature published in this area); finally, the data is presented for an external audience for dissemination of the results. The survey was influenced by an interpretivism research philosophy, and as a result, the emphasis was placed on incorporating open-ended questions into the survey to garner respondents' perspectives. This philosophy sought to investigate the insider's view, or real meanings, of the CSR/SD area, with respondents who are considered to have an excellent social knowledge of the domain (Wahyuni 2012). The cross-tabulation method was utilised to quantitatively analyse the relationships between the multiple variables (range of factors) that were identified in the study.

7.3.3   Validity and Reliability

Validity and reliability of the questionnaire tool were ensured by initially developing the questions from the literature concerning CSR and then ensuring that each question investigated that which it set out to examine. The questions were worded carefully, and one or more questions were used in the analysis of specific individual themes for investigation. Particular attention was given to questions regarding enablers and barriers to businesses progressing from a low level of CSR engagement, i.e. doing the minimum to comply with regulations, to upper levels where CSR is fully integrated with the overall business strategy. It is claimed by Bini et al. (2018) that it is at the top level that companies effectively incorporate sustainability strategies into their business model, i.e. its value creation process. Before administering the survey, an initial pilot questionnaire was sent to several CSR practitioners and academics to ensure the validity and reliability of the tool used. The pilot study was sent to two academic professors in Madrid, Spain, and one CSR funding manager in the UK to obtain feedback on each question and evaluate how the respondents might interpret the questions' meaning.

7.4   Results/Discussion

Following the completion of all the questionnaires, the data was downloaded to the researchers PC for data analysis. Questions are broadly categorised as background/profile details, educational qualifications, experience, managerial level, leadership/CSR training, advantages of sustainability reporting, disadvantages of sustainability reporting, the role of religion in business development, government input into CSR. These are the range of factors considered for analysis. The quantitative data are presented in the form of descriptive statistics. Descriptive statistics summarise patterns in the responses of the cases in the sample. Descriptive analysis is conducted and presented in this research in both graphical and tabular form. The background/profile data is presented in Sect. 7.4.1. Subsequently, the results of the thematic analysis on the qualitative text data from the open-ended questions are presented in Sect. 7.4.2. The cross-tabulation method of data analysis is carried out on the categorical data to quantitatively analyse the relationships between multiple variables, e.g. the likelihood that senior managers who replied to the survey perceive their business to be operating at the upper, most active, CSR level, compared to middle managers who hold the same view.

7.4.1 Profile of Respondents (Quantitative Data)

Table 7.1 'Other' category (18.6% of respondents stated other industries)

- Services,
- Real Estate and Renting and Leasing, Agriculture, Forestry, and Construction,
- Sustainability,
- Chemicals,
- Consumer Electronics,
- Professional Services,
- Aquaculture,
- Primary/Secondary Meat Processing,
- Agri-Food,
- We operate in Horticulture, and also have a Juice and cider manufacturing operation,
- Food production, Manufacturing, Sales, Innovation/Product development,
- Marketing, export and sales services,
- Food,
- Food processing,
- Consultancy,
- CSR.

Table 7.1 Industries in which survey respondents organisations operate

Q. In which industry do you operate?

		Frequency	Percent
Valid	Other (please specify)	16	18.6
	Agriculture, Forestry, Fishing	14	16.3
	Mining, Quarrying, and Oil and Gas extraction	3	3.5
	Utilities	5	5.8
	Construction	3	3.5
	Manufacturing	25	29.1
	Wholesale	1	1.2
	Retail	3	3.5
	Transportation and Warehousing	1	1.2
	Finance and Insurance	5	5.8
	Real estate and Renting and leasing	3	3.5
	Waste management	1	1.2
	Educational services	2	2.3
	Health care	3	3.5
	Public administration	1	1.2
	Total	86	100

Manufacturing businesses, at 29%, have the highest percentage of all the industrial sectors to have responded to the survey, as seen in Table 7.1. Torugsa et al. (2012) concur with this higher level of CSR activity among the manufacturing industry in SME's in Australia. Other sectors that appear to have significant CSR focus, according to company listings on sustainability databases and who contributed to this research are Agriculture/Forestry/Fishing (16%), Utilities (6%) and Finance/Insurance (6%). Research into industry-specific CSR activities indicates that the industries type and sectors significantly affect CSR activities. Policies and regulations are designed for specific industries, which means that companies operating under a particular umbrella face similar restrictions (O'Connor and Shumate 2010) such as environmental or workforce regulations. To this end, certain companies face more stringent regulations which may account for the high number of responses from companies in the manufacturing industry as shown in Table 7.1. Dabic et al. (2016) identified a high level of CSR activities and reporting from companies whose operations were closer to consumers in the value chain (e.g. speciality retailers) as well as those industries utilising environmental resources or heavy industry that may impact on the environment.

There is an even spread of company size in terms of the number of people employed by the company, as shown in Table 7.2. The respondents include small businesses, Small-Medium-Enterprises (SME), large industries and Multi-National-Corporations (MNC's). O'Connor et al. (2017) found a difference in CSR priorities depending on different company size (SME and MNC).

Of the 81 respondents to the management level question, it was noticeable that only 24 respondents (30%) operated at top management level. The management level at which most respondents worked, 43 in total, was the middle management level (53%). The remaining 14 respondents (17%) stated that they were neither of these categories but were advisors, CSR consultants, project managers, analysists, and business owners. In the UN/BCG 2014 Report, only 20% of the 3795 senior

Table 7.2 Size of company in terms of employee numbers

Q. How many employees does your organisation have?		Frequency	Percent
Valid	Between 1 and 9	10	11.6
	Between 10 and 49	15	17.4
	Between 50 and 249	14	16.3
	Between 250 and 999	14	16.3
	Between 1000 and 9999	18	20.9
	Between 10,000 and 100,000	13	15.1
	Greater than 100,000	1	1.2
	Total	85	98.8
Missing	System	1	1.2
Total		86	100

Table 7.3 Levels of CSR engagement

Q. Which of the following levels of engagement best describes your Corporate Social Responsibility/Sustainable Development activities within your organisation?

		Frequency	Percent	Valid percent	Cumulative percent
Valid	Upper level	51	59.3	66.2	66.2
	Middle level	17	19.8	22.1	88.3
	Lower level	7	8.1	9.1	97.4
	None of the above	2	2.3	2.6	100
	Total	77	89.5	100	
Missing	System	9	10.5		
Total		86	100		

and middle managers interviewed believed that their senior-level executive boards provide substantial insight on sustainability issues.

Thirty-two of the 86 respondents (41%) had between 1 year and 5 years' experience in a CSR role and 11 (14%) had less than one years' experience. These values may indicate that either (i) CSR has a renewed prominence within businesses as 55% of participants had 5 years' experience, or lower, in their CSR role or (ii) newer members of staff are delegated this role. While the earliest seminal piece of CSR literature was produced by Bowen (1953) and publications in the area increased rapidly through the following decades (Carroll 1999), there is now huge interest and demands upon CSR operations and reporting within business, industry and manufacturing globally (Birkey et al. 2017). Recent high-profile financial and corporate ignominies of which the global public are cognisant maintains firm focus upon CSR issues (Habek and Wolniak 2013; Perez 2015).

Respondents were asked about the level of CSR engagement (integration of their CSR strategy/business strategy) at which they perceived their company to be operating. Of the 77 respondents to this question, the majority of these (51) see themselves as engaging on the upper level of CSR activities, as shown in Table 7.3. Cross-tabulation revealed that 75% of senior managers saw their business to be operating at the top level of CSR engagement (Table 7.3). In comparison, there were a much lower percentage of (respondent) middle managers, 58%, who perceived their business to be operating at the upper CSR levels. It is at the top level of CSR engagement that companies effectively integrate sustainability strategies into their business model, i.e. its value creation process (Bini et al. 2018). There are many possible factors that influence the level of CSR engagement (Hu et al. 2018) among which is the type of business ownership and whether the business is listed on the Stock Exchange. It is possible that this finding may reflect what Dingwerth and Eichinger (2010) identify as an 'opportunity to enhance a firm's reputation aspect' of CSR reporting and that senior managers may identify any opportunity even a research study to highlight their firm positively. Conversely, it may be as claimed by Birkey et al. (2017) that increasingly senior managers are embracing CSR because of its impact on shareholders and the future earnings response coefficient (FERC). The high number of respondents who perceive themselves to be operating in the upper level may reflect the global move towards sustainability reporting (Medel-Gonzalez et al. 2013; Habek and Wolniak 2013; Sartori et al. 2017). Nine respondents did not answer this question (Missing). These respondent companies were selected based on national (Irish) and international sustainability reporting databases so, therefore, the low number (7 out of 77) on the lower level (businesses who do the minimum to comply with regulations) was predictable. All of the companies have their sustainability measures, and endeavours embedded on their websites and promote themselves as being responsible corporate citizens similar to findings in the literature (Li et al. 2013; Dingwerth and Eichinger 2010). Website content is reproduced (anonymously) and contextually analysed for four sample companies who responded to the survey, the results can be viewed in Appendix 7H.

Question 6 requested the respondents to detail the sustainability reporting framework to which their business is aligned. Most of the Irish-based respondent companies were aligned to the Origin Green reporting framework, and the majority of the UK and USA companies were aligned to the GRI reporting framework. A brief overview of these two main frameworks is given in the 'Appendix 7I'.

Of the 66 respondents to the question regarding the highest academic qualification two had PhD qualifications, 28 were educated to Masters level, and equally, 28 respondents had a primary degree. A total of 7 respondents had diploma level qualifications, and one had certificate level qualifications. Twenty people did not answer this question. In their study of university students (mainly those of law, economics and social sciences) Harring and Jagers (2018) claim that higher education students opinions regarding climate change and environmental issues appear to be influenced by personal belief and background as well as trust in government and its institutions. Commenting on higher education and qualifications and understanding of environmental issues and policy, Harring and Jagers (2018) state that individuals with university degrees would be deemed to have heightened ability to discern and critically analyse issues such as climate change, however equally in their study, they found varying degrees of support for formal environmental policy initiatives. Shephard et al. (2015) found little effect of higher education upon changes in environmental behaviours over time (Table 7.4).

The majority of the 65 respondents to this question, 31 (48%) completed their academic training in Ireland. The two respondents who selected 'Other' completed their academic training in the UK, France and Australia. Twenty-one respondents choose not to answer this question (Missing). There was an even spread of the time period for their most recent academic training by the respondents—17 within the last two years (26.2%), 15 between two years and 5 years ago (23.1%), 17 between 5 and 10 years ago (26.2%), and 16 more than ten years ago (24.6%). Twenty-one respondents skipped this question. This means that 75.5% of respondents completed their most recent academic training within the last ten years. Harring and Jagers (2018) claim that graduate opinions may reflect somewhat the philosophy of their specific educational institution. The opinions regarding CSR expressed by almost 50% of this study's respondents may be influenced by perceptions and state policy interventions in environmental issues within a particular Irish context (Table 7.5).

Table 7.4 Location of academic training of survey respondents

Q. Where did you complete your business academic training?

		Frequency	Percent	Valid percent	Cumulative percent
Valid	Other country (please specify)	2	2.3	3.1	3.1
	United Kingdom	21	24.4	32.3	35.4
	Ireland	31	36	47.7	83.1
	Europe (excluding UK and Ireland)	4	4.7	6.2	89.2
	United States of America	7	8.1	10.8	100
	Total	65	75.6	100	
Missing	System	21	24.4		
Total		86	100		

Table 7.5 Leadership module on academic programmes for survey respondents?

Q. As part of your academic training, did you complete modules on 'leadership'?

		Frequency	Percent	Valid percent	Cumulative percent
Valid	Yes	34	39.5	53.1	53.1
	No	30	34.9	46.9	100
	Total	64	74.4	100	
Missing	System	22	25.6		
Total		86	100		

Of the 64 respondents to this question, 34 stated that they had taken leadership modules as part of their academic programme but almost as much again, 30, said that their academic training did not include leadership modules. Surprisingly, 22 (over a quarter of the respondents) did not answer this question. Leadership was identified in the literature as being very important for championing CSR issues within businesses (Pless et al. 2012; Waldman 2011). Utilising the cross-tabulation method to quantitatively analyse the relationship between the management level at which the respondent operates in their business *and* whether there were leadership modules on their academic training programmes, 62.5% of top (senior) level managers indicated that they had completed leadership modules as part of their educational programme while only 32.6% of middle managers stated that they had completed leadership modules on their educational programmes. Andrejczuk (2016) recognised the need for strong leadership as a primary driver to responsible business development. Senior managers are in a position to assign resources to CSR activities as part of a company's strategy (Hutchison-Krupat and Kavadias 2015) and therefore have an ideal opportunity to show leadership in this area. More than half (53%) of respondents, those with CSR responsibilities, stated that they operated at 'middle-management'

Table 7.6 Geographical areas for 'leadership' modules on academic programmes

		Leadership modules on academic programme	
Academic training in		Yes (%)	No (%)
Ireland		65	35
UK		37	63
USA		57	43
Europe (Exc Irl & UK)		50	50

level within their organisation (Sect. 7.4.1). Research by Beliveau (2013) and Kealy (2015) found that the primary role of a middle manager was to implement a deliberate strategy decided on by senior, strategic level, managers. It would appear that managers at the middle level find it difficult to enact change and champion a specific facet like CSR within the culture of the organisation. Waldman et al. (2006) claim that the CEO's vision of CSR may impact on their lower-level managers' view of CSR in their decision-making process.

Graduates who completed their academic training in Ireland are more likely (65%) to have undertaken leadership modules as part of their educational degree programmes based on the results shown in Table 7.6.

Of the 64 respondents to the question of completing a CSR/Sustainable Development module as part of their academic training, 36 replied that they had not done so (56.3% in Table 7.7). Twenty-eight respondents stated that they had done so (43.8%), and 22 people skipped this question. None of the USA respondents had completed modules on CSR/SD as part of their academic programmes. If the 22 respondents who did not answer this question are taken into account, then only one-third of respondents (32.6%) had completed CSR modules on their academic programmes. While completing CSR modules as part of academic programmes is only one tool by which CSR is better understood, the figures suggest a missed opportunity for CSR proliferation.

There were 60 respondents (out of 86) to the question of the potential for a faith-based input as a positive influence on sustainable business development. The majority of these (46 or 76.7%) stated NO while 14 respondents (23.3%) stated YES. There

Table 7.7 CSR modules on academic programmes for survey respondents?

Q. As part of your academic training, did you complete modules on 'Corporate Social Responsibility/Sustainable Development'?

		Frequency	Percent	Valid percent	Cumulative percent
Valid	Yes	28	32.6	43.8	43.8
	No	36	41.9	56.3	100
	Total	64	74.4	100	
Missing	System	22	25.6		
Total		86	100		

was a significant number (26) of respondents who skipped this question. While some respondents felt that a faith-based input (such as Catholic Social Teaching) could be '*very helpful when combined with an overall programme in ethics/philosophy*' (Respondent #3, 7C), others suggested '*it could be positive or negative*' (Respondent #13, 7C). Another issue that emerged was the possibility of risk in introducing faith into business practices. The risk is that '*it would be difficult to work with faith-based organisations that are non-affiliated and inclusive enough to meet the needs of our diverse stakeholders*' (Respondent #1, 7C), sometimes covering many countries and cultures for global multinational businesses. One of the respondents claimed that '*the tenets of the Judeo-Christian religion that have actually formed most western thoughts and actions, and in a business context they can be a powerful force for good, positive social and environmental behaviour*' (Respondent #11, 7C). Pope Francis (2015) claims that a combination of religion and science can benefit each other as solutions to our current environmental, and human, degradation is sought. In a review of literature from 30 countries on the role of religion in firm-level innovation, Assouad and Parboteeah (2018) found that religion may transmit positive values such as expectations, rules, and boundaries that guide and influence workplace behaviour. They asserted that their findings hold true for any religion, e.g. Christian, Buddhism, and Muslim (Assouad and Parboteeah 2018). However, the authors acknowledged that they took a benevolent view of religion while Chan-Serafin et al. (2013) highlighted the fact that religion has the potential to be a force for good *or* bad in an organisation.

Five respondents contributed additional comments at the end of the survey, where CEO commitment and CSR skills and training were part of the pertinent issues specified.

7.4.2 Open-Ended Questions (Qualitative Data)

7.4.2.1 Advantages Associated with Sustainability Reporting

The respondents identified many issues associated with sustainability reporting, these included Stakeholder Engagement (external) [SE]—Improvement [IM]—Regulation [R]—Human (internal) [H]—Ethics [E]—Strategy [S]—Top Management [TM]—Marketing [M]. A total of 41 respondents contributed to this question. Crosstabulation analysis identified the most pertinent issues that respondents associated with sustainability reporting within their organisations. The advantageous issues that arose are listed as follows:

- Improvement [IM] 18/41
- Stakeholder Engagement [SE] 15/41
- Human (internal) [H] 12/41
- Strategy [S] 10/41
- Regulation [R] 10/41

- Ethics [E] 7/41
- Marketing [M] 5/41
- Top Management [TM] 2/41

The data from respondents revealed that the most common advantageous issue that was associated with a business implementing a sustainability reporting process was the fact that the business could use the values in the report to improve performance in future years, *'provides ongoing data for which continuous improvement plans can be designed and measured'* (Respondent #1, 7B) (18 respondents out of a total of 41 mentioned this when replying to this question). De Villiers et al. (2016) concur, and claim that there are potential advantages, financial and non-financial, associated with reporting on sustainability issues within their business.

Over half of the respondents who claimed that the sustainability reporting process might assist in the reduction of their carbon footprint had worked as CSR managers for periods between 5 and 10 years. The remaining respondents with similar response were employed as CSR managers for periods between 1 and 5 years. It was also stated that there was more stakeholder engagement due to the reporting process, *'way of communicating with our stakeholders about our values'* (Respondent #14, 7B). The completion of the sustainability report requires input from many stakeholders and the more stakeholders that become involved in the process, potentially the more benefits there is accrued by the business. Cai et al. (2012) concurred with this view and found benefits of CSR activity and subsequent reporting for the stakeholders within even the less popular 'sinful' industries of tobacco, alcohol, biotech, and weaponry. This positive aspect includes both internal, *'supports employee retention/recruitment'* (Respondent #4, 7B) and external stakeholders, *'more of our customers are looking for products produced responsibly'* (Respondent #34, 7B).

Further comments on the advantages of CSR reporting from respondent #21 (recent PhD graduate) included the recognition that businesses with an influential CSR culture can use the ethos for *'alignment with business development strategy'.* Respondent #21 (Appendix 7B) operates as a top manager in a company that has between 10,000 and 100,000 employees. This company operates on the upper level of CSR engagement. Nuttaneeya et al. (2012) concurred with these opinions identifying the link between CSR activities and positive strategic advantages not only in larger companies but in their particular review of Small to Medium Enterprises (SME's).

Other opinions on the advantages of CSR activity and reporting was that CSR *'promotes and encourages an ethical ethos throughout the company'* and *'fosters good corporate governance'* (Respondent #30, 7B). Lee et al. (2013) identified similar positive perceptions of employees, in their study on employee attitude to implementing and reporting CSR activities and how it may influence employee satisfaction and commitment to an organisation. However, Rodrigo and Arenas (2008) found concerning employee attitudes towards CSR, there always exists the indifferent and even dissenting employee within an organisation. They found, in their study confined to specific construction firms, that concerning employee opinion and attitude towards CSR there appeared to be two main opinions, a healthy support attitude or relative indifference but unusually no cautious attitude 'wait and see' approach (Rodrigo and Arenas 2008).

CSR *'helps with B2B sales'* (Respondent #41, 7B). Respondent #41 (Appendix 7B) is the owner/manager in an Irish-based family manufacturing business with between 1 and 9 employees and who has less than one years' experience in CSR. They state that the business operates at the middle level of the CSR/business strategy integration spectrum and appreciate the support that Bord Bia (Appendix 7H) affords them in their CSR endeavours. It would appear that CSR is more easily accepted by employees when business benefits are seen to follow on as a result of deciding to implement CSR strategies and initiatives. Soundararajan et al. (2018) concur with this view that the voices of other related stakeholders, e.g. employees, customers, suppliers and communities must be taken into consideration to advance understanding and benefits of CSR.

7.4.2.2 Disadvantages Associated with Sustainability Reporting

Bekefi and Epstein (2016) claim that there are potential disadvantages to embracing sustainability reporting within a business. In-depth analysis of the data for this current study identified *time* as one of the disadvantages to sustainability reporting and this is shown in Fig. 7.2. The word 'time' was subjected to a text search and shows how respondents described some of the disadvantages associated with sustainability reporting and how they are explicitly linked to the issue of 'time'.

The disadvantage of committing a company resource (time) to complete the sustainability report (usually annually) was one of three company resource factors seen as a disadvantage to sustainability reporting, the other two being a human resource and a financial resource. The full list of issues that developed as a result of the coding process for this 'disadvantages' part of the question is as follows:

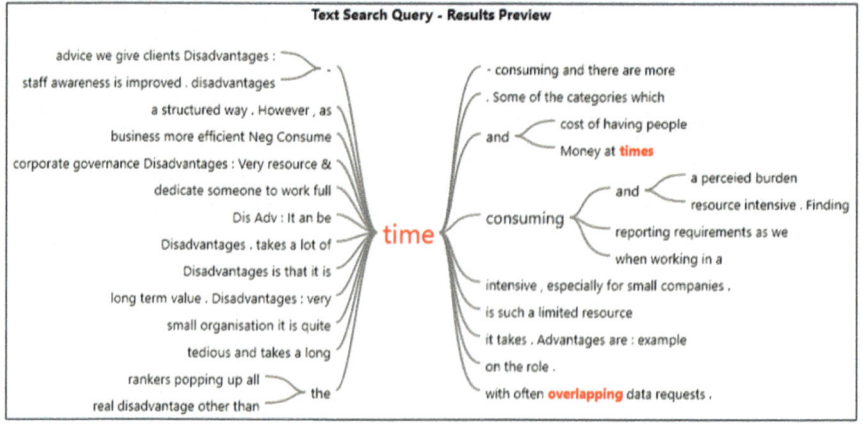

Fig. 7.2 Word tree with 'time' as the subject identifying disadvantages to CSR reporting

- Non-Standard Reporting [NS] (7 respondents out of a total of 28 mentioned this in their response: 7/28)
- Surface Effects [SF] (3/28)
- Resources (Time) [RST] (15/28)
- Resources (Human) [RSH] (3/28)
- Resources (Financial) [RSF] (3/28)
- Complexity [CX] (5/28)
- Measurement [M] (2/28)

The disadvantage of not aligning sufficient resources to implement the sustainability reporting process effectively is a major issue identified in this study, *'time-consuming reporting requirements as we do not have someone solely hired for environmental/CSR/Sustainable Development'* (Respondent #42, 7B). Bini et al. (2018) supported this view declaring that if CSR is to be a strategic issue within a business, then it is undoubtedly incumbent of top (senior-level) managers to align sufficient resources to do the job properly This includes a more significant portion of company time, human, and financial resources allocated to CSR-related activities, *'very resource and time intensive, especially for small companies'* (Respondent #30, 7B).

The lack of a cohesive approach to sustainability reporting highlighted in the introduction chapter by Xu et al. (2016) (Sect. 2.2.1) was evident in the responses to this part of the survey also, *'the lack of a structured approach to data gathering in line with a sustainability framework means it is more challenging'* (Respondent #26, 7B). The *'cost and complexity and a lack of common standards within and between industries'* were identified as disadvantages to sustainability reporting (Respondent #5, 7B).

A cross-tabulation method of analysis was conducted to analyse whether respondents who had completed CSR modules on their academic programmes also stated that the measuring and reporting process presented difficulties as a disadvantage to reporting on sustainability issues. Nine out of 36 respondents (25%) who had NOT completed CSR modules as part of their academic programmes mentioned the measuring and reporting process in their replies while 11 out of 28 respondents (39%) who HAD completed CSR modules also mentioned measuring and reporting as an issue. Four of the respondents who had completed CSR modules stated that they could save on energy costs and reduce their carbon footprint while only one respondent who had not completed CSR modules stated likewise. Within the educational sphere, in a study of Indian and Mauritius managerial students Saxena and Mishra (2017) declared that 'Overall, CSR assumes centre stage for achieving successful outcomes in the long-term aspects and cannot be underestimated'. Interestingly in a study of student diversity and attitudes towards CSR within the educational field Haski-Leventhal et al. (2017) identified female students and more mature students as having a more positive attitude towards CSR and responsible management education.

7.4.2.3   Level of Support for CSR Duties

Of the 54 respondents to the question of support for CSR duties, 23 stated that they had positive, top management, support as they carried out their CSR duties, *'we have the full support from the Board and the Executive'* (Respondent #17, 7D). Only 6 out of the 54 respondents had a low level of support. Some respondents stated that they would receive support if the CSR activities were deemed to be profitable, *'support is generally available if the sustainability team can be seen to be adding value and not just introducing something that does not add anything to the business'* (Respondent #12, 7D). Respondent #12 works in a UK-based real estate company with between 250 and 999 employees and operates at middle manager level. Respondent #32 also stated that support for CSR increases as verifiable resource reductions are achieved. Respondent #32 works for an Irish-based meat processing company at middle manager level. Many of the non-respondents to this question were employed as consultants and educators. Reimer et al. (2018) found that the interaction between the CEO and the Top Management Team (TMT) is an important aspect in defining a firms' CSR profile while Sanzo et al. (2012) identified that top management support for innovation affects environmental management policies and the firm missions' link to sustainability.

7.4.2.4   Governmental Involvement in Promoting Sustainable Business Development

There were 50 respondents to the question as to whether governments should be more actively involved in promoting Sustainable Business Development and the vast majority (90%) stated YES and a minority (10%) stated NO. The respondents were encouraged to submit other comments on what level of governmental involvement if any, they wish to see in assisting businesses in the CSR/Sustainable Development area. Comments were made that *'governments can and should legislate to prevent the most negative impacts a company may have on society, e.g. environmental'* (Respondent #2, 7E). Governments can also *'help to set the minimum standards expected and provide avenues/grants/regulation to encourage leadership and innovation'* (Respondent #5, 7E). There was also a suggestion that *'sustainable business development should be a careful balance of Public/Private Partnership'* (Respondent #11, 7E). The importance and relevance of the government role identified here concur with conclusions made by Steurer (2010) who declare that 'CSR started out as a neo-liberal concept that helped to downscale government regulations, but that it has in turn matured into a more progressive approach of societal co-regulation'. Interestingly Skare and Golja (2014) found that countries that had government policy strongly supporting CSR activity appeared to achieve higher growth rates, while Kealy (2014) also found that effective government policy is one of the essential inputs required for sustainable business development (Fig. 3.2 in Sect. 3.5.3). Many of the respondents believed that governments could do more in terms of enacting legislation, particularly in the area of environmental protection measures. They perceived that this would encourage all companies to become more efficient with their resources. Almatrooshi et al. (2018) argue that there is scope for governments to play a more significant role in mediating the relationship between the business community and civil society. The study by Almatrooshi et al. (2018) did focus solely on the United Arab Emirates (UAE) public/private communities but the suggestions made are consistent with the findings in this study.

7.4.2.5 Barriers to Integrating CSR and Business Strategy

The main issues that emerged as barriers to a business integrating CSR into its overall business strategy were identified as follows:

- CSR is Expensive [EX] (20/50)
- No Stakeholder Priority [NS] (15/50)
- Education [ED] (8/50)
- Time Resource [TM] (8/50)
- Difficult to Measure [DM] (6/50)
- Lack of Leadership [LL] (5/50)
- Profit Driven [PD] (4/50)
- Lack of Governance [LG] (1/50)

Prior published literature in the CSR barriers theme identified key barriers to the implementation of CSR in businesses in Poland (Andrejczuk 2016). Two of the key barriers to the implementation of CSR were (i) the belief that CSR 'does not pay off' and brings no benefits and (ii) a lack of education among managerial staff (Andrejczuk 2016). These two findings by Andrejczuk (2016) concur with the findings of this study namely 'CSR is Expensive' (20 respondents out of a total of 50 mentioned this) and 'Education' (8 respondents). The findings also concur with the findings in a study by Hussain et al. (2018) who state that a shortage of skills, a lack of training and lack of knowledge are barriers to social sustainability in a healthcare supply chain (Hussain et al. 2018, Table 6). Two of the enablers identified by Hussain et al. (2018) were listed as (i) resource efficiency and (ii) information sharing (Hussain et al. 2018, Table 5). The majority of respondents (95%) who stated that company resources were a barrier to CSR advancement worked in small companies who had a workforce not exceeding 1000 personnel. Interestingly a similar view regarding the resources barrier to CSR advancement was expressed by only 5% of respondents from the larger organisations, i.e. those with a workforce exceeding 1000 personnel. Of the total number of respondents to this question about company size in which the respondent worked, 62% belonged to the smaller company size, i.e. less than 1000 while 38% worked in the larger companies (Table 7.2). A sample of the responses included '*often the sustainable way is often quite expensive to implement or slightly more expensive than the alternative, it is also more time expensive*' (Respondent #45, 7F). One of the barriers was stated as '*a perception that sustainability is expensive and does not add value*' (Respondent #11, 7F). Respondent #11

(7F) is a degree-level educated middle manager in a UK-based real estate company who suggests that sustainability has the potential to become part of the everyday business if we can move away from 'perceptions' and prove that sustainability can 'actually' add value. Another barrier was a *'lack of understanding of the breadth of sustainability'* (Respondent #7, 7F). Barriers included a *'lack of senior management buy-in'* (Respondent #22, 7F) and a *'lack of clear leadership, vision, mission and non-aligned values'* (Respondent #25, 7F). Respondent #25 (Appendix 7F) is a recent PhD graduate (completed their PhD within the last two years) who also states that one of the disadvantages to CSR reporting is that *'accountants usually don't see the benefits'*. The measurement aspect of CSR was stated as a barrier because of a *'lack of metrics to measure impact of activities across all areas, including social'* (Respondent #31, 7F). Respondent #9 (Appendix 7B) a middle manager in a small construction company (between 50 and 249 employees) who operate at the upper level, claims that *'collating the information and measuring the impact of the report'* is a disadvantage to the sustainability reporting process. The difficulty of measuring the impact of CSR was expressed by respondents at all levels of management in this study. Many studies in the literature also identify this CSR accurate measurement issue (Gjolberg 2009; Diego Paredes-Gazquez et al. 2016; Suárez-Cebador et al. 2018). Skarmeas and Leonidou (2013) go further and declare that many people doubt the proficiency of CSR measurement and reporting while more recently Etter et al. (2019) declares that now social media have enabled the ordinary citizen outside an organisation's normative shareholder group, to bypass the traditional gatekeeping function of business evaluators, and they can now proceed autonomously to make and highlight individual judgments regarding corporative activity, through social media outlets.

7.5 Conclusions

7.5.1 Respondent Profile

A general profile of the respondents based on the responses (Sect. 7.4.1) to the survey can be identified as follows; the respondents mainly work in the manufacturing industry and are employed within SME's. They operate at the middle-management level and have less than ten years' experience in the CSR area. They perceive their business to be operating on the upper level of the CSR/strategy integration paradigm. They are educated to graduate and postgraduate level, academic training completed within the last ten years. Within their education, only half of the respondents are likely to have completed modules on leadership and less than half likely to have completed CSR modules as part of their academic training.

7.5.2 Main Themes Identified to Enhance CSR/Business Strategy Integration

As the research progressed, two main themes emerged regarding the employees' opinions on the issues that allowed the enhancement of CSR/business strategy integration. The respondents identified a certain level of difficulty in measuring and reporting the economic, environmental and social aspects of their management activities, particularly the non-financial activities, i.e. their environmental and social aspects. On the ground, this *'level of difficulty'* appears to be affected and augmented by an insufficient amount of time being allotted to the task, in particular the time resource issue was a difficulty for the majority of the respondents who were employed in small companies, i.e. less than 1000 employees. The comments by participants identifying disadvantages to sustainability reporting (Sect. 7.4.2.2) suggest that there is a lack of a common universal standard (framework) within and between industries by which they can report on the three elements of CSR (Fig. 7.1). The need for an improved standard reporting framework was identified by respondents at both middle manager level (Respondent #26, 7B) and also at top manager level (Respondent #5, 7B). Respondent #26 worked for an agri-food company that employs between 1000 and 9999 employees and operated at the lower level of CSR engagement while respondent #5 worked at a manufacturing company that employs between 10,000 and 100,000 employees who operated at the middle level of CSR engagement. Respondent #26 (Appendix 7B) uses the Origin Green reporting framework, while Respondent #5 (Appendix 7B) uses the GRI reporting framework (Appendix 7J). While the GRI is a well-established framework, it is clear that completing the report is not a trivial task. The fact that the GRI implementation help manual is 269 pages long gives some indication of the complexity of the task. Indeed, some of the respondents to the current study identified the *'complexity'* problem (Respondent #5, Appendix 7B) about the disadvantages associated with sustainability reporting (Sect. 7.4.2.2). Vigneau et al. (2015) concur, and highlights this need for greater coordination between the various sustainability standards to increase the potential to improve corporate accountability and CSR reporting. Respondent #26 (Appendix 7B) is a Managing Director of a small company (between 10 and 49 employees) which operates on the upper level of CSR engagement and she identified the fact that a *'lack of a structured approach to data gathering in line with a sustainability framework means it is more challenging to develop data-driven reporting'* as a disadvantage to the current sustainability reporting process. Andrejczuk (2016) agrees, citing that a structured approach to data gathering may assist businesses in ascertaining whether indeed their CSR initiatives were 'paying off'. In a Poland-based study by Andrejczuk (2016, Table 7) 45.1% of respondents identified the 'belief that CSR does not pay off and brings no benefits' as a key barrier to the implementation of CSR in businesses. A robust data gathering measurement stage appears to be of the utmost importance to inform whether CSR initiatives are paying off, rather than inconclusive beliefs that may or may not be true (Diego Paredes-Gazquez et al. 2016). The robust data measurement stage may help to alleviate the opinion by a large number of respondents to the on-line survey that

CSR is expensive (Sect. 7.4.2.5). CSR initiatives, by definition, should impact positively on the profit bottom line, the environmental bottom line, and the people bottom line (Elkington 1997). A new, coherent, standardised measuring and reporting CSR framework would surely help CSR to have closer *'alignment with business development strategy'* (Respondent #21, Appendix 7B). Respondent #21 (Appendix 7B) is a PhD educated top manager in a large (between 10,000 and 100,000 employees) manufacturing company who operate at the upper level of CSR engagement.

The second main theme identified in this study was the *education/knowledge* theme. It was found that 56.3% of people who responded to the question claimed that they did not complete modules in CSR/Sustainable Development as part of their degree programme (Table 7.7). While it is acknowledged that there are other interfaces than academic degree programmes and industry, (e.g. research, consultancy; Iyer 2014) it is anticipated that embedding CSR topics on degree programmes would have a positive impact on CSR activity and outcomes (Gatti et al. 2019). The lack of CSR-based modules is not in line with the UN/GRLI guidelines on mission, strategy, and activities within business schools set up to develop the next generation of globally responsible leaders. Considering that the GRLI was formed in 2004 and most participants (75.5%) to this study completed their academic training within the last ten years (Sect. 7.4.1), it is disappointing that only 43.8% of respondents to the question completed CSR modules as part of their academic programmes (Table 7.7). While the GRLI academic initiative is a step in the right direction, there appears to be a lack of policing of the UN/BCG initiative to convince more academic institutions to participate in propagating CSR modules into their degree and master's programmes. A study by Biermann et al. (2017) found that the academic support for the integration of the social, environmental, and economic dimensions of the United Nations 17 SDG up to 2030 is of critical importance. A UN/BCG (2014) report claims that one of the most substantial barriers to greater executive-level engagement is a lack of sustainability expertise among board members (not explicitly examined in this current study) and the Global Corporate Sustainability Report (2013) where 63% of the respondents ranked education as the top global sustainability challenge (Sect. 3.2.1). This finding is also in agreement with a study by Smith and Ronnegard (2016) who claim that the main focus in business schools is on producing graduates who, as future managers, make decisions to further the interests of the shareholders first and foremost. An earlier study by Lamsa et al. (2008) also found that students from a business school in Finland valued the shareholder model ahead of the broader spectrum stakeholder model. However, a more recent article by Deer and Zarestky (2017) recommends practical tools for business schools to enhance lecturing on CSR modules. It appears from the literature that more recent research is highlighting the need for the social and environmental bottom lines in business education and not focusing solely on the profit bottom line. Interestingly some of the findings of this current study indicated that CSR would be supported in business *if* CSR initiatives were found to be financially profitable (Sect. 7.4.2.3). One of the main barriers to CSR/strategy integration was that respondents believed that CSR is expensive (Sect. 7.4.2.5) and some businesses felt they could not afford to invest in initiatives, namely as yet unproven beneficial environmental initiatives.

There were some uncertain opinions expressed around whether CSR was adding value to company activities (Respondent #11, Appendix 7F) and also on how best to measure the impact of a company's CSR activities (Respondent #9, Appendix 7B). It was considered that these issues might be alleviated by better education in CSR. Indeed it is also clear that the development and promulgation of robust measuring and evaluation tools for CSR initiatives is now imperative not only to avoid green-washing but to build the body of knowledge within academia. It does now appear that specialist research is emerging around CSR measurement within the diverse business world (Perez and Rodrigues del Bosque 2013; Wells et al. 2015). The positives and negatives of CSR initiatives need to be highlighted to adopt sustainable processes and avoid uncertain investments within the CSR umbrella, as found by Kealy (2017). The research by Kealy (2017) involved the conduction of an empirical evaluation of a 300-kW wind turbine initiative. The study found that the investment did not reduce the cost of the electricity bill for the SME despite the €280,000 capital investment. The SME undertook this project as part of its CSR program, hoping to fulfil its CSR obligations and introduce some cost-saving measures for the company. On the surface it appeared a positive initiative all round, however following more in-depth analysis and measurement of the project, there were minimal (at best) cost savings and minimal CO_2 emissions reduction due to this wind turbine investment decisions (Kealy 2017). It was clear that the robust analysis and measurement of the CSR initiative adopted in this case revealed a different outcome to the stakeholders than they initially envisaged or indeed realised.

7.6 Management Implications

In conclusion, this study sought to identified vital enablers and barriers encountered by businesses in their efforts to integrate CSR within the overall business strategy. Data were obtained by on-line survey and analysed using the thematic analysis method. It is essential to highlight that the two main enablers/barriers were identified across a range of different demographics in the study, e.g. company size, managerial level, or industry type. Central themes were established as follows:

- Measurement of CSR outcomes needs to be intensified and promulgated. This would ensure a body of robust empirical data on CSR activity and outcomes would exist, which would be accessible to businesses, and by which CSR investment decisions can be evaluated. The CSR reporting process needs to be more coherent and standardised than it is currently.
- Academic institutional programmes and interdisciplinary research should be encouraged to include more CSR/SD modules; these should be across all business, engineering, and economic disciplines. Those out in the field measuring CSR activities should be encouraged not only to submit data as part of regulatory and legislative obligations but also to document, publish and disseminate findings

within not only academia but also throughout the business/industrial community, as a reference point for CSR activities and initiatives in the field.

Unless the two key enablers/barriers are addressed, i.e. measurement/reporting of the outcomes of CSR initiatives *and* education/knowledge of CSR issues, it would appear that many businesses may struggle to integrate their CSR strategy successfully into its business model, i.e. its value creation process (Bini et al. 2018).

7.7   Limitations and Further Research

The businesses who participated in this research were selected based on their residing on sustainability-reporting databases (Origin Green and GRI). This research does not assess the many businesses that do not seek to be part of a sustainability reporting instrument but appear to be operating as socially responsible companies. Further research is needed to assess companies that decide not to seek inclusion on sustainability-reporting databases.

References

Adams, C. A., & Evans, R. (2004). Accountability completeness, credibility and the audit expectations gap. *Journal of Corporate Citizenship, 14,* 97–115.

Ali, S. H. (2016, August). Reconciling Islamic ethics, fossil fuel dependence, and climate change in the middle east. *Review of Middle East Studies, 50*(2), 172–178.

Almatrooshi, S., Hussain, M., Ajmat, M., & Tehsin, M. (2018). Role of public policies in promoting CSR: Empirical evidence from business and civil society of UAE. *Corporate Governance: The International Journal of Business in Society, 18*(6), 1107–1123. https://doi.org/10.1108/CG-08-2017-0175.

Andrejczuk, M. (2016). *The development of CSR in Poland as seen by managers* (Research Papers of Wroclaw University of Economics, nr 423). https://doi.org/10.15611/pn.2016.423.01.

Aspling, A. (2013). Business, management education, and leadership for the common good. In D. L. Everett (Ed.), *Shaping the future of business education: Relevance, rigor and life preparation* (pp. 40–58). New York: Palgrave Macmillan.

Assouad, A., & Parboteeah, K. P. (2018). Religion and innovation, a country institutional approach. *Journal of Management, Spirituality & Religion, 15*(1), 20–37. https://doi.org/10.1080/14766086.2017.1378589.

Bekefi, T., & Epstein, M. J. (2016, November). 21st century sustainability. *Strategic Finance, 98*(11), 28–37.

Beliveau, J. (2013). Middle Managers' role in transferring person-centred management and care. *Service Industry Journal, 33*(13/14), 1345–1362.

Biermann, F., Kanie, N., & Kim, R. E. (2017, June). Global governance by goal setting: The novel approach of the UN sustainable development goals. *Current Opinion in Environmental Sustainability, 26–27,* 26–31. https://doi.org/10.1016/j.cosust.2017.01.010.

Bini, L., Bellucci, M., & Giunta, F. (2018, January). Integrating sustainability in business model disclosure: Evidence from the UK mining industry. *Journal of Cleaner Production, 171,* 1161–1170. https://doi.org/10.1016/j.jclepro.2017.09.282.

Birkey, R. N., Guidry, R. P., & Patten, D. M. (2017, December). Does CSR reporting really impact FERC's? *Accounting and the Public Interest, 17*(1), 144–162.

Bowen, H. (1953). *Social responsibility of the businessman*. New York: Harper & Row.

Braun, V., & Clarke, V. (2006). Using thematic analysis in psychology. *Qualitative Research in Psychology, 3*(2), 77–101.

Bruntland Commission Report. (1987). *World commission on environment and development, our common future*. New York, NY: Oxford University Press.

Cai, Y., Jo, H., & Pan, C. (2012, July). Doing well while doing bad? CSR in controversial industry sectors. *Journal of Business Ethics, 108*(4), 467–480. https://doi.org/10.1007/s.10551-011-1103-7.

Carollo, L., & Guerci, M. (2017). Between continuity and change: CSR managers' occupational rhetorics. *Journal of Organisational Change Management, 30*(4), 632–646. https://doi.org/10.1108/JOCM-05-2016-0073.

Carroll, A. B. (1979). A three-dimensional model of corporate performance. *Academic Management Review, 4*(4), 497–505.

Carroll, A. B. (1999, September). Corporate social responsibility: Evolution of a definitional construct. *Business & Society, 38*(3), 268–295.

Carroll, A. B., & Shabana, K. M. (2010). The business case for corporate social responsibility: A review of concepts, research and practice. *International Journal of Management Reviews, 12*(1), 85–105. https://doi.org/10.1111/j.1468-2370.2009.00275.x.

Catalin, C., & Nicoleta, R. (2011). International biomass trade and sustainable development: An overview. *Annals of the University of Oredea, Economic Science Series, 20*(2), 47–54.

Chan-Serafin, S., Brief, A. P., & George, J. M. (2013, September–October). How does religion matter and why? *Organisation Science, 24*(5), 1291–1600. https://doi.org/10.1287/orse.1120.0797.

Chen, C.-J., Guo, R.-S., Hsiao, Y.-C., & Chen, K.-L. (2018, November). How business strategy in non-financial firms moderates the curvilinear effects of corporate social responsibility and irresponsibility on corporate financial performance. *Journal of Business Ethics, 92*, 154–167. https://doi.org/10.1016/j.jbusres.2018.07.030.

Cubas-Diaz, M., & Martinez Sedano, M. A. (2018, October). Measures for sustainable investment decisions and business strategy—A Triple Bottom Line Approach. *Business Strategy and the Environment, 27*(1), 16–38. https://doi.org/10.1002/bse.1980.

Dabic, M., Colovic, A., Lamotte, O., Painter-Morland, M., & Brozovic, S. (2016). Industry-specific CSR: Analysis of 20 years of research. *European Business Review, 28*(3), 250–273. https://doi.org/10.1108/EBR-06-2015-0058.

De Villiers, C., Rouse, P., & Kerr, J. (2016, November). A new conceptual model of influences driving sustainability based on case evidence of the integration of corporate sustainability management control and reporting. *Journal of Cleaner Production, 13*, Part A, 78–85. https://doi.org/10.1016/j.jclepro.2016.01.107.

de Vries, G., Terwel, B. W., Ellemers, N., & Daamen, D. D. L. (2015, May). Sustainability or profitability? How communicated motives for environmental policy affect public perceptions of corporate greenwashing. *Corporate Social Responsibility & Environmental Management, 22*(3), 142–154. https://doi.org/10.1002/csr.1327.

Deer, S., & Zarestky, J. (2017). Balancing profit and people: Corporate social responsibility in business education. *Journal of Management Education, 41*(5), 727–749. https://doi.org/10.1177/1052562917719918.

Diego Paredes-Gazquez, J., Miguel Rodriguez-Fernandez, J., & de la Cuesta-Gonzalez, M. (2016). Measuring corporate social responsibility using composite indices: Mission impossible? The case of the electricity utility industry. *Spanish Accounting Review, 19*(2), 142–153. https://doi.org/10.1016/j.rcsar.2015.10.001.

Dingwerth, K., & Eichinger, M. (2010, August). Tamed transparency: How information disclosure under the Global Reporting Initiative fails to empower. *Global Environmental Politics, 10*(3), 74–96.

Elkington, J. (1997). *Cannibal with forks: The Triple Bottom Line of 21st century business.* Capstone: Gabriola Island, BC.

Engert, S., Rauter, R., & Baumgartner, R. J. (2016, January). Exploring the integration of corporate sustainability into strategic management: A literature review. *Journal of Cleaner Production, 112*, Part 4, 2833–2850. https://doi.org/10.1016/j.jclepro.2015.08.031.

Etter, M., Ravasi, D., & Colleoni, E. (2019, January). Social media and the formation of organisational reputation. *Academy of Management Review, 44*(1), 28–52. https://doi.org/10.5465/amr.2014.0280.

European Foundation for Management Development. (2014). Available from: http://www.efmd.org/. Accessed 17 September 2014.

Fortanier, F., Kolk, A., & Pinkse, J. (2011, October). Harmonisation in CSR reporting: MNE's and Global CSR standards. *Management International Review, 51*(5), 665–696. https://doi.org/10.1007/s11575-011-0089-9.

Gatti, L., Ulrich, M., & Seele, P. (2019, January). Education for sustainable development through business simulation games: An exploratory study of sustainable gamification and its effects on students learning outcomes. *Journal of Cleaner Production, 207*, 667–678. https://doi.org/10.1016/j.jclepro.2018.09.130.

Geographical. (2015, October). Call to action. *Geographical Magazine Ltd, 87*(10), 9.

Gjolberg, M. (2009, March). Measuring the immeasurable? Constructing an index of CSR practices and CSR performance in 20 countries. *Scandinavian Journal of Management, 25*(1), 10–22. https://doi.org/10.1016/j.scaman.2008.10.003.

Globally Responsible Leadership Initiative. (2014). Available from: http://www.grli.org/. Accessed 17 September 2014.

Gobbels, M. (2002). Reframing corporate social responsibility: The contemporary conception of a fuzzy notion. *Journal of Business Ethics, 44*, 95–105.

Gray, R. (2010). Is accounting for sustainability actually accounting for sustainability and how would we know? An exploration of narratives of organisations and the planet. *Accounting, Organisations and Society, 35*, 47–52.

Habek, P., & Wolniak, R. (2013). Analysis of approaches to CSR reporting in selected European Union countries. *International Journal of Economic Research, 416*, 79–95.

Hack, L., Kenyon, A. J., & Wood, E. H. (2014). A Critical Corporate Social Responsibility (CSR) timeline: How should it be understood now? *International Journal of Management Cases, 16*(4), 46–55.

Halati, A., & He, Y. (2018, July). Intersection of economic and environmental goals of sustainable development initiatives. *Journal of Cleaner Production, 189*, 813–829. https://doi.org/10.1016/j.jclepro.2018.03.323.

Harring, N., & Jagers, S. C. (2018). Why do people accept environmental policies? The prospects of higher education and changes in norms, beliefs and policy preferences. *Environmental Education Research, 24*(6), 791–806. https://doi.org/10.1080/13504622.2017.1343281.

Haski-Leventhal, D., Pournader, M., & McKinnon, A. (2017, November). The role of gender and age in business students' values, CSR attitudes, and responsible management education: Learnings from the PRME international study. *Journal of Business Ethics, 146*(1), 219–239. https://doi.org/10.1007/s10551-05-2936-2.

Holme, L., & Watts, P. (2000). *Corporate social responsibility; making good business sense.* Geneva: World Business Council for Business Development.

Hu, Y. Y., Zhu, Y., Tucker, J., & Hu, Y. (2018). Ownership influence and CSR disclosure in China. *Accounting Research Journal, 31*(1), 8–21. https://doi.org/10.1108/ARJ-01-2017-0011.

Hussain, M., Ajmal, M. M., Gunasekaran, A., & Khan, M. (2018, December). Exploration of social sustainability in healthcare supply chain. *Journal of Cleaner Production, 203*, 977–989. https://doi.org/10.1016/j.jclepro.2018.08.157.

Hutchison-Krupat, J., & Kavadias, S. (2015, February). Strategic resource allocation: Top-down, Bottom-up, and the value of strategic buckets. *Management Science, 61*(2), 391–412. https://doi.org/10.1287/mnsc.2013.1861.

Isil, O., & Hernke, M. T. (2017). The Triple-Bottom-Line: A critical review from a transdisciplinary perspective. *Business Strategy and the Environment, 26*, 1235–1251. https://doi.org/10.1002/bse.1982.

Iyer, T. (2014, March). Role of industry-academia interface for filling the skill gap. *CLEAR International Journal of Research in Commerce & Management, 5*(3), 52–54.

Jones Christensen, L., Peirce, E., Hartman, L. P., Hoffman, W. M., & Carrier, J. (2007, July). Ethics, CSR, and sustainability education in the financial times top 50 global business schools: Baseline data and future research directions. *Journal of Business Ethics, 73*(4), 347–388.

Jones, P., Comfort, D., & Hiller, D. (2011). Sustainability in the global shop window. *International Journal of Retail & Distribution Management, 39*(4), 256–271. https://doi.org/10.1108/09590551111117536.

Kealy, T. (2014, October). Sustainable business development: An Irish perspective. *International Journal of Humanities and Social Science, 4*(12), 166–179.

Kealy, T. (2015, January) Do middle managers contribute to their organisation's strategy? *International Journal of Humanities and Social Science, 5*(1), 108–116, ISSN: 2220-8488 (Print). http://www.ijhssnet.com/journal/index/2920.

Kealy, T. (2017, February). Stakeholder outcomes in a wind turbine investment; is the Irish energy policy effective in reducing GHG emissions by promoting small-scale embedded turbines in SME's? *Renewable Energy, 101*, 1157–1168. https://doi.org/10.1016/j.renene.2016.10.007.

Lamsa, A. M., Vehkapera, M., Puttonen, T., & Pesonen, H. L. (2008). Effects of business education on women and men students' attitudes on corporate responsibility in society. *Journal of Business Ethics, 82*, 45–58. https://doi.org/10.1007/s10551-007-9561-7.

Lee, M.-D. P. (2008). A review of the theories of corporate social responsibility: Its evolutionary path and the road ahead. *International Journal of Management Reviews, 10*(1), 53–73. https://doi.org/10.1111/j.1468-2370.2007.00226.x.

Lee, E. M., Park, S.-Y., & Lee, H. J. (2013, October). Employee perception of CSR activities: Its antecedents and consequences. *Journal of Business Research, 66*(10), 1716–1724. https://doi.org/10.1016/j.jbusres.2012.11.008.

Li, N., Hilgard, J., Scheufele, D. A., Winneg, K. M., & Jamieson, K. H. (2016, December). Cross-pressuring consevative Catholics? Effects of Pope Francis's encyclical on the US public opinion on climate change. *Climatic Change, 139*(3–4), 367–380. https://doi.org/10.1007/s10584-016-1821-z.

Li, Y., Zhang, J., & Foo, C.-T. (2013). Towards a theory of social responsibility reporting: Empirical analysis of 613 CSR reports by listed corporations in China. *Chinese Management Studies, 7*(4), 519–534. https://doi.org/10.1108/CMS-09-2013-0167.

Luthra, S., Mangla, S. K., Chan, F. T. S., & Venkatesh, V. G. (2018). Evaluating the drivers to information and communication technology for effective sustainability initiatives in supply chains. *International Journal of Information Technology & Decision-Making, 17*(1), 311–338. https://doi.org/10.1142/S0219622017500419.

Medel-Gonzalez, F., Garcia-Avila, L., Acosta-Beltran, A., & Hernandez, C. (2013). Measuring and evaluating business sustainability: Development and application of corporate index of sustainability performance. *Sustainability Appraisal: Quantitative Methods and Mathematical Techniques for Environmental Performance Evaluation*, 33–61. https://doi.org/10.1007/978-3-642-3208-1_3.

Medina Rivilla, A., & Medina Dominguez, C. (2014). Competencies, education and sustainable development: A case study. *Economic Insights—Trends and Challenges, 66*(1), 25–34.

Milne, M. J., & Gray, R. (2013). 'W(h)ither Ecology? The Triple Bottom Line, the Global Reporting Initiative, and Corporate Sustainability Reporting. *The Journal of Business Ethics, 118*, 13–29.

Norman, W., & MacDonald, C. (2004). Getting to the bottom of the 'Triple-Bottom-Line'. *Business Ethics Quarterly, 14*(2), 243–262.

Nowell, L. S., Norris, J. M., White, D. E., & Moules, N. J. (2017). Thematic analysis: Striving to meet the Trustworthiness Criteria. *International Journal of Qualitative Methods, 16*, pp. 1–13. https://doi.org/10.1177/1609406917733847.

Nuttaneeya, A. T., O'Donohue, W., & Hecker, R. (2012, September). Capabilities, proactive CSR and financial performance in SME's: Empirical evidence from an Australian manufacturing industry sector. *Journal of Business Ethics, 109*(4), 483–500. https://doi.org/10.1007/s.10551-011-1141-1.

O'Connor, A., Parcha, J. M., & Tulibaski, K. L. G. (2017). The institutionalisation of corporate social responsibility communication. *Management Communication Quarterly, 31*(4), 503–532. https://doi.org/10.1177/0893318917704512.

O'Connor, A., & Shumate, M. (2010, July). An economic industry and institutional level of analysis of corporate social responsibility communication. *Management Communication Quarterly, 24*(4), 529–551.

Ozgur, I., & Hernke, M. T. (2017). The Triple Bottom Line: A critical review from a transdisciplinary perspective. *Business Strategy and the Environment, 26*(8), 1235–1251. https://doi.org/10.1002/bse.1982.

Perez, A. (2015). Corporate reputation and CSR reporting to stakeholders: Gaps in the literature and future lines of research. *Corporate Communications: An International Journal, 20*(1), 11–29. https://doi.org/10.1108/CCIJ-01-2014-0003.

Perez, A., & Rodriguez del Bosque, I. (2013, December). Measuring CSR image: Three studies to develop and to validate a reliable measurement tool. *Journal of Business Ethics, 118*(2), 265–286. https://doi.org/10.1007/s10551-012-1588-8.

Pless, N. M., Maak, T., & Waldman, D. A. (2012, November). Different approaches toward doing the right thing: Mapping the responsibility orientations of leaders. *Academy of Management Perspectives, 26*(4), 51–65.

Pope Francis. (2015). Laudato Si'. *Encyclical letter of the Holy Father Francis on care for our common home*. Veritas.

Rambaud, A., & Richard, J. (2015, December). The "Triple Depreciation Line" instead of the "Triple Bottom Line": Towards a Genuine Integrated Reporting. *Critical Perspectives on Accounting, 33*, 92–116. https://doi.org/10.1016/j.cpa.2015.01.012.

Reimer, M., Van Doorn, S., & Heyden, M. L. M. (2018, September). Unpacking functional experience complementarities in senior leaders' influence on CSR strategy: A CEO—Top management team approach. *Journal of Business Ethics, 151*(4), 977–995. https://doi.org/10.1007/s10551-017-3657-5.

Rodrigo, P., & Arenas, D. (2008, December). Do employees care about CSR programs? A typology of employees according to their attitudes. *Journal of Business Ethics, 83*(2), 265–283. https://doi.org/10.1007/s10551-007-9618-7.

Sandbu, M. (2015, September 15). Critics question success of UN's millennium development goals. *Financial Times*, Special Reports.

Sanzo, M. J., Alvarez, L. I., Rey, M., & Garcia, N. (2012, September). Perceptions of top management commitment to innovation and R&D marketing relationship effectiveness: Do they affect CSR? *Annals of Public & Cooperative Economics, 83*(3), 383–405. https://doi.org/10.1111/j.1467-8292.2012.00468.x.

Sartori, S., Witjes, S., & Campos, L. M. S. (2017, December). Sustainability performance for Brazilian electricity power industry: An assessment integrating social, economic, and environmental issues. *Energy Policy, 111*, 41–51. https://doi.org/10.1016/j.enpol.2017.08.054.

Savitz, A. (2012). *The Triple Bottom Line: How today's best-run companies are achieving economic, social and environmental success—And how you can too*. San Francisco, USA: Wiley.

Saxena, M., & Mishra, D. K. (2017). CSR perception: A global opportunity in management education. *Industrial & Commercial Training, 49*(5), 231–244. https://doi.org/10.1108/ICT-12-296-0085.

Schwerhoff, G., & Sy, M. (2017, August). Financing renewable energy in Africa—Key challenges of the sustainable development goals. *Renewable and Sustainable Energy Reviews, 75*, 393–401. https://doi.org/10.1016/j.rser.2016.11.004.

Seto-Pamies, D., & Papaoikonomou, E. (2016, July). A multi-level perspective for the integration of Ethics, Corporate Social Responsibility and Sustainability (ECSRS) in management education. *Journal of Business Ethics, 136*(3), 523–538.

Shephard, K., Harraway, J., Lovelock, B., Mirosa, M., Skeaff, S., Slooten, L., et al. (2015, February). Seeking learning outcomes appropriate for 'education for sustainable development' and for higher education. *Assessment & Evaluation in Higher Education, 40*(6), 855–866. https://doi.org/10.1080/02602938.2015.1009871.

Silberhorn, D., & Warren, R. C. (2007). Defining corporate social responsibility: A view from big companies in Germany and the UK. *European Business Review, 19*(5), 352–372. https://doi.org/10.1108/09565340710818950.

Skare, M., & Golja, T. (2014, May–June). The impact of government CSR supporting policies on economic growth. *Journal of Policy Modeling, 36*(3), 562–577. https://doi.org/10.1016/j.jpolmod.2014.01.008.

Skarmeas, D., & Leonidou, C. N. (2013, October). When consumers doubt, watch out! The role of CSR Skepticism. *Journal of Business Research, 66*(10), 1831–1838. https://doi.org/10.1016/j.jbusres.2013.02.004.

Smith, W. K. (2014). Dynamic decision-making: A model of senior leaders managing strategic paradoxes. *Academy of Management Journal, 57*(6), 1592–1623. https://doi.org/10.5465/amj.2011.0932.

Smith, N., & Ronnegard, D. (2016, March). Shareholder primacy, corporate social responsibility, and the role of business schools. *Journal of Business Ethics, 134*(3), 463–478.

Soundararajan, V., Jamali, D., & Spence, L. J. (2018). Small business social responsibility: A critical multilevel review, synthesis and research agenda. *International Journal of Management Reviews, 20*, 934–956. https://doi.org/10.1111/ijmr.12171.

Sridhar, K., & Jones, G. (2013, January). The three fundamental criticisms of the triple bottom line approach: An empirical study to link sustainability reports in companies based in the Asia-Pacific region and TBL shortcomings. *Asian Journal of Business Ethics, 2*(1), 91–111. https://doi.org/10.1007/s13520-012-0019-3.

Steurer, R. (2010, March). The role of governments in corporate social responsibility: Characterising public policies on CSR in Europe. *Policy Sciences, 33*(1), 49–72. https://doi.org/10.1007/s11077-009-9084-4.

Suarez-Cebador, M., Rubio-Romero, J. C., Pinto-Contreiras, J., & Gemar, G. (2018, September/October). A model to measure sustainable development in the hotel industry: A comparative study. *Corporate Social Responsibility and Environmental Management, 25*(5), 722–732. https://doi.org/10.1002/csr.1489.

Tang, Z., Hull, C. E., & Rothenberg, S. (2012). How corporate social responsibility engagement strategy moderates the CSR-financial performance relationship. *Journal of Management Studies, 49*(7), 1274–1303. https://doi.org/10.1111/j.1467-6486.2012.01068.x.

Taylor, J., Vithayathil, J., & Yim, D. (2018, September). Are Corporate Social Responsibility (CSR) initiatives such as sustainable development and environmental policies value enhancing or window dressing? *Corporate Social Responsibility and Environmental Management, 25*(5), 971–980. https://doi.org/10.1002/csr.1513.

Torugsa, N. A., O'Donoghue, W., & Hecker, R. (2012, September). Capabilities, proactive CSR and financial performance in SME's: Empirical evidence from an Australian manufacturing industry sector. *Journal of Business Ethics, 109*(4), 483–500. https://doi.org/10.1007/s10551-011-1141-1.

United Nations Global Compact, in collaboration with BCG. (2014, January). Joining forces: Collaboration and leadership for sustainability. *MIT Sloan Management Review*, Research Report 2015.

United Nations—Global Corporate Sustainability Report. (2013). Available from: http://www.unglobalcompact.org/AboutTheGC/global_corporate_sustainability_report.html. Accessed 24 February 2014.

Vaismoradi, M., Jones, J., Turunen, H., & Snelgrove, S. (2016). Theme development in qualitative content analysis and thematic analysis. *Journal of Nursing Education and Practice, 6*(5), 100–110. https://doi.org/10.5430/jnep.v6n5p100.

Venturelli, A., Caputo, F., Leopizzi, R., Mastroleo, G., & Mio, C. (2017, January). How can CSR identity be evaluated? A pilot study using a fuzzy expert system. *Journal of Cleaner Production, 141*, 1000–1010. https://doi.org/10.1016/j.jclepro.2016.09.172.

Vigneau, L., Humphreys, M., & Moon, J. (2015, October). How do firms comply with international sustainability standards? processes and consequences of adopting the Global Reporting Initiative. *Journal of Business Ethics, 131*(2), 469–486. https://doi.org/10.1007/s10551-014-2278-5.

Wahyuni, D. (2012). The research design maze: Understanding paradigms, cases, methods and methodologies. *Journal of Applied Management Accounting Research, 10*(1), 69–80.

Waldman, D. A. (2011). Moving forward with the concept of responsible leadership: Three key caveats to guide theory and research. *Journal of Business Ethics, 98,* 75–83.

Waldman, D. A., Siegel, D. S., & Javidan, M. (2006, December). Components of CEO transformational leadership and corporate social responsibility. *Journal of Management Studies, 43*(8), 1703–1725.

Wells, V. K., Manika, D., Gregory-Smith, D., Tahen, B., & McCowlen, C. (2015, June). Heritage tourism, CSR and the role of employee environmental behaviour. *Tourism Management, 48,* 399–413. https://doi.org/10.1016/j.tourman.2014.12.015.

Xu, J., Li, L., & Zheng, B. (2016, June). Wind energy generation technological paradigm diffusion. *Renewable and Sustainable Energy Reviews, 59,* 436–449.

Yin, R. K. (2008). *Case study research: Design and methods* (4th ed.). London: Sage.

Zivkovic, J. (2012). Strengths and weaknesses of business research methodologies: Two disparate case studies. *Business Studies Journal, 4*(2), 91–99.

Chapter 8
Findings

Abstract This chapter discusses the main findings associated with the research study. The main findings are focused on companies who decide to invest in wind turbine technologies as part of their environmentally friendly initiatives and need to evaluate those decisions. Methods to assess such decisions robustly are recounted. Results from the 10-kW, 300-kW, and 3.5-MW wind turbine initiatives are discussed considering the novel evaluation framework. Wind turbine economic and environmental outcomes are presented and are found to be disappointing. By TBL definition, if economic and environmental sustainability outcomes are not attained, the social aspect of sustainability is also not accomplished. There appears to be a groupthink consensus that the installation of wind turbines guarantees sustainability success, an idea that is tested in this current book. The measurement of sustainability outcomes and an increase in sustainability/CSR knowledge are identified as key enabling factors in the attempts to more closely integrate corporate sustainability strategy with the overall business strategy.

8.1 Introduction

This chapter highlights the principal findings of the overall research book publication. The results are presented and then discussed through three critical stages (topics) as the framework is development:

- Wind turbines through an SD/CSR lens (Sect. 8.2.1),
- Wind turbines: A technical and economic evaluation (Sect. 8.2.2),
- The Key enabling and inhibiting factors in integrating sustainability/CSR strategy with business strategy (Sect. 8.2.3).

Subsequently, a unique framework which evolved from this study is developed and presented (Fig. 8.3 in Sect. 8.2.4). The framework is specifically an evaluation tool for managing SD/CSR environmental/wind turbine project decisions but can be adapted to evaluate any sustainability decisions.

© Springer Nature Switzerland AG 2020   179
T. Kealy, *Evaluating Sustainable Development and Corporate Social Responsibility Projects*, https://doi.org/10.1007/978-3-030-38673-3_8

8.2 Key Findings

8.2.1 Wind Turbines Through an SD/CSR Lens

According to a literature review into Corporate Social Responsibility (CSR), the term remains an ambiguous concept (Lau et al. 2018). Reasons for this ambiguity were found to include (i) CSR crosses into several academic disciplines (ii) it is not adequately taught in educational institutions (iii) globalisation has challenged the roles and responsibilities of employees, corporations, and the state. However, CSR and its related concept Sustainable Business Development (and the shortened version Sustainable Development, SD) is primarily considered to be concerned with business decision-making that considers economic, environmental, and social factors. Of these three distinct, yet interconnected, factors, this study found that there is a substantial body of research assessing the relationship between CSR and financial (economic) business performance and CSR and human (social) aspects, but there are minimal published research studies considering the CSR/environmental connection. Wind turbines are considered part of the environmental dimension of CSR as it is anticipated that their use in generating electricity helps to reduce CO_2 emissions (Cullen 2013). Research on this understudied wind turbine/CSR aspect is limited, and the literature significantly appears to indicate a lack of integrity or universality. Some studies claim that this gap in published literature may be due to the lack of a systematic analytical framework, a gap addressed in this current publication.

An ethnographic study by Kealy (2014b) probed the opinions of what constitutes sustainable development practices in Ireland using qualitative thematic analysis on data obtained from semi-structured interviews. Summary of the findings confirmed that the three well established vital constituents must be considered if businesses are to be developed in a sustainable way, namely *Profit, Planet,* and *People* are still valid. The research found that business leaders guided by a robust moral/ethical compass who intently consider the three components when making business decisions can develop their business sustainably. The three parts were deemed to be interconnected, so success in one aspect brings benefits to the other two (synergies and interplay demonstrated in Table 3.1). SD/CSR initiatives that are not cost-effective or environmentally beneficial do not assist society to develop sustainably.

8.2.2 Wind Turbines: A Technical and Economic Evaluation

Research informs us that wind turbines are considered part of the environmental sector (of the Three-P's, Elkington 1997) as they generate carbon-neutral electrical energy and therefore reducing our dependence on CO_2 emitting fossil-fuel-driven electrical generators. The 10-kW (Kealy 2014a) and 300-kW (Kealy 2017) wind turbine case studies (Chaps. 4 and 6, respectively) found that there was a minimal financial benefit due to the investment decisions and minimal CO_2 emission reductions. Actual objective measured data was obtained from the turbines by the connection of electrical power/energy meters on the turbine output cables. Analogue electrical energy meters (RDM) were designed and installed safely by this researcher (Fig. A6.4) and the method provided triangulation of energy output data with the (standard) energy output data provided by the digital indicators. The digital energy measurement meter utilises a signal (current, voltage) sampling method, while the analogue energy measurement meter utilises a torque method for energy measurement. This method is the first time such a triangulation of energy data was presented and advanced the rigour and robustness of these findings. In addition to the energy measurements, the turbine output power quality was also measured using the three-phase Fluke 1735 Power Logger Analysist. It was found that there were continuous (short-term) ramping variations in the power output signal from the 300-kW DFIG generator shown in Figs. 6.3 to 6.6 and the power output from the 10-kW three-phase synchronous generator shown in Fig. 4.3. The study on the 3.5-MW wind farm (Kealy et al. 2015) allowed data to be retrieved and analysed from the on-site digital energy meters, but access to the data required to evaluate the energy benchmarks was not forthcoming as the data was deemed to be commercially sensitive by the data owners. Further research is needed in this case study to determine if the 9,808,318 kWh units of electricity produced by the turbines per year *offset* the same amount that would have been generated by (mainly) fossil-fuel-driven generators. This enquiry can be achieved by gaining access to the number of yearly energy units imported through both the 38-kV Supply A and 38-kV Supply B from the national electricity grid in Fig. 5.3 which should have reduced by the same amount. Disappointingly the significant number of wind turbine installations connected to the grid as stated in Sect. 1.1.3 did not show any significant improvements to the national electricity generation benchmark values as shown in Table 1.2 (and Table A4.1). This disappointing result contrasts with the good news story of Ireland's reduction in energy import dependency from 88% in 2015 to 69% in 2016 as a result of the Corrib gas field coming on stream.

The triangulation method is shown in Fig. 8.1 using on-site captured objective measured energy and power quality data produced by the (10-kW and 300-kW) wind turbines augmented the credibility and validity of the novel closed-loop framework. Also, the robustness of the new framework was increased by the fact that there were two different measuring techniques used to measure the number of alleged kWh units of electrical energy produced by the turbines namely an analogue energy measurement tool and a digital energy measurement tool. This reliable validated data was used for the completion of the company sustainability reports where accurate energy benchmarking was then instigated. Research by the Centre for Sustainability and Excellence (CSE 2017) claimed that companies who report well-using goals and externally assured performance data see that their efforts are reflected in their financial performance. A shortage of objective wind turbine data also raised significant ethical and moral issues as companies selling turbines promised purported positive outcomes which it appears is only achievable on a modelled basis but in practical terms would never produce the promised result either financially or technically.

Fig. 8.1 Triangulation of objective measured energy and power quality data

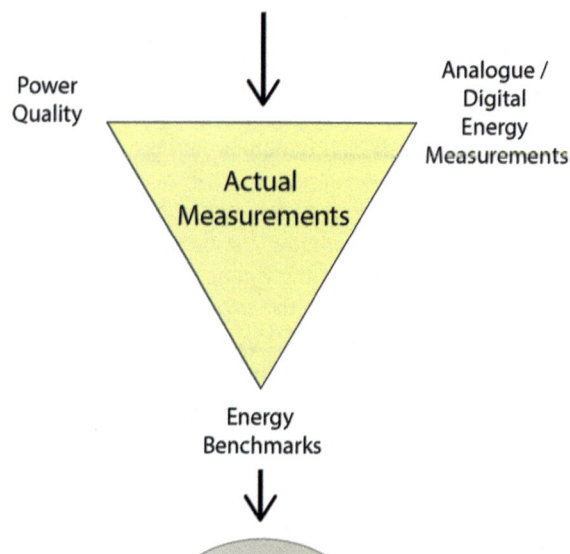

Power Quality

Analogue / Digital Energy Measurements

Actual Measurements

Energy Benchmarks

There is a dearth of prior actual measured evidence by which to go on, and no previously established norms have been established. This book contributes to the body of knowledge by measuring and presenting actual turbine power output data in line with the new closed-loop SD/CSR evaluation model. Most companies state turbine power and energy outputs based on modelled data or data under ideal test conditions which may or may not be accurate. It may be that the unpredictable nature of wind and its half-second-by-half-second variation in real terms (as measured and demonstrated in the ramping constantly phenomenon in Figs. 6.3 to 6.6) would appear not have been factored in accurately during a modelled predicted power output stage. Cullen (2013) argues that generators operate more efficiently when operating in a steady manner near maximum capacity and emission rates can change during ramping periods. There was a significant number of ramping (up and down) periods of wind generator output according to the data shown in Figs. 6.3 to 6.6. The wind turbine installations examined in this book are connected in parallel (embedded) with the conventional power generators, and therefore, the wind turbine output variation affects the running conditions of the conventional power generators. When the wind turbine output drops, the conventional generator (normally gas turbines) must be able to respond rapidly to the demand change. When fossil-fuel-driven generators are quickly ramped up and down (to compensate for the ramping of the parallel-connected wind turbines), their fuel use (and thus CO_2 emissions) may be larger than when they are operating at a steady power level (Katzenstein and Apt 2009). One explanation for the virtually static energy benchmark values (Table 1.2 and Table A4.1) is that traditional (fossil-fuel) generators are unable to cope with the constantly ramping tendency associated with parallel-connected wind generators. They are expected to take up the slack but are unable to do so. Chapters 4 and 6 are examples of published research where specific wind turbine installations failed to meet promised expectations. It was clear from the results of the in-depth case study carried out on the 300-kW wind turbine and presented in Chap. 6 that the negative results were not recognised by any of the stakeholders involved in the project, i.e. government, turbine supplier, utility provider, business owner, business employees. It was only when a second electrical energy meter was installed by this researcher that the error was detected (Fig. A6.4). A multi-disciplinary CSR-focused education would not have prevented the errors or improved the wind turbine investment outcomes, but it is likely to assist in the methods of robustly identifying the red flag issues regarding turbine investment decisions. There may be a culture of groupthink in Ireland currently whereby the installation of wind turbines is unquestioned as having a positive value, economically, environmentally, and socially. To counteract this groupthink phenomenon, businesses who have invested in wind turbine technology such as those presented in Chaps. 4–6 possess a corporate social opportunity in informing other businesses of the difficulties they encountered in the life-time of their wind energy project. The ineffective results from the wind turbine case study presented in Chaps. 4 and 6 do not appear to be an isolated phenomenon as Melia (2017) reports that Ireland is the worst-performing country in Europe concerning acting on climate change. Melia's report (2017) is based on the 2018 Climate Change Performance Index published in November 2017 at the UN climate talks in Bonn, Germany. There appears to be a large gulf between Irish government rhetoric and the reality of effective government policy implementation.

8.2.3 Wind Turbines: The Key Enabling Factors in Integrating SD/CSR and Business Strategy

Following on from the technical and economic evaluation of the wind turbine projects, some essential SD/CSR enabling factors were identified that allow a dynamic integration of CSR with the overall business strategy. The enabling factors were evident in many of the previous case studies but were the primary focus of the research presented in Chap. 7. These following factors are identified as critical enablers for ensuring that CSR is better integrated with the overall business strategy:

- The measuring of the actual outputs of SD/CSR initiatives and reporting on the results, specifically concerning the results from the non-financial components of CSR, requires collaboration from academia and business practice to develop a standardised, coherent framework by which to guide the process. One such input to the new framework using objective measuring techniques is presented in Sect. 6.3.2. In Sects. 3.2.5, 4.3.4, and 6.5.1.1, reporting of such data is discussed.
- The need for an increase in the quantity of CSR/SD education at third-level (and other) academic institutions. In data presented in Table 7.7, 56% of respondents did not complete CSR/SD modules as part of their educational programmes. In Sect. 4.5, the investor who spent €26,620 on the wind turbine had limited technical

Fig. 8.2 Identifying key
CSR enabling factors in
integrating CSR/business
strategy

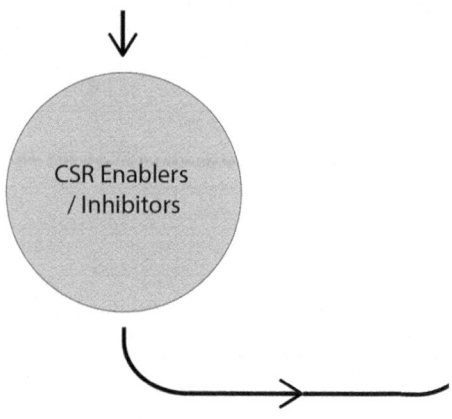

understanding of the venture; instead, they relied on using a 'gut feeling' in making the investment decision. There should be further educational opportunities to upskill before making such significant financial investments. The SD/CSR education should be interdisciplinary (business, engineering, and economic disciplines) (Fig. 8.2).

As demonstrated in Sect. 7.4.2.2, many businesses argue that the process of reporting on their non-financial aspects of business activity (human and environmental) is time-consuming and difficult. It may be advantageous to include high-level company employees (such as the Chief Financial Officer) in the completion of the sustainability report that is intended to highlight the importance and value of such a task (Herve 2018).

The dearth of multi-disciplinary (business, engineering, and economic) knowledge/education regarding aspects of wind turbine investments was a common theme running through the findings of this research study (Chaps. 4 and 6). Uncertainty surrounded the realistic expectations with regards to the benefits (economic, CO_2 reduction) of the 10-kW, 300-kW and 3.5-MW turbine installations. Instead, they were influenced perhaps by a consensus that significant benefits would follow as a result of such investments. This general consensus-seeking behaviour without enough robust empirical data can evolve and morph into a groupthink mentality. Janis (1983) argued that groupthink was as a result of deterioration of mental efficiency and moral judgement and that group pressures can force people towards conformity. There possibly exists group pressure and tension within the energy sector in Ireland currently as the country is in a race against time to meet binding EU energy targets. The shortage of empirical data to inform the industry and assist in more of a robust objective decision-making process may have been a factor in pressurising the stakeholders in wind energy initiatives into a negative groupthink mentality leading to tremendous support (see Table 2.2 and Sect. 2.5.4) for the, as yet unproven, positive benefits of wind energy projects (Kelman et al. 2017). Education in energy benchmarking would allow stakeholders to possess the knowledge to be capable of identifying aspects of a project that determines a good, or bad, wind turbine investment. This current study can contribute to a more informed decision-making process with a specific focus on wind energy investments and can assist the business to know the practical steps that need to be implemented to robustly critique such investments thereby advancing their value-maximising choices (Benabou 2013).

Findings from this study suggest that the relatively new business activities related to renewable energy technologies would indicate that many companies selling turbines, solar, and other renewable energy technologies in some cases do not possess the necessary qualifications or expertise to do so. This failing was especially significant in the case studies presented in Chap. 4 (Kealy 2014a) and Chap. 6 (Kealy 2017) and raised some significant ethical issues. This researcher recommends that an escalation in modules focused on CSR in third-level academic institutions would help to improve this situation. The turbine traders promised energy output values that did not meet those guaranteed values. In Sect. 7.4.2.5, many respondents stated that education (or lack of knowledge in the understanding of CSR) is a barrier to businesses enhancing a CSR culture existing and flourishing within an organisation.

While the effectiveness of the UN in promoting a world order of responsible business activity is questionable (Sect. 3.6), many business personnel who responded to the online survey (Chap. 7) wish to see government (local, national, international) has more involvement in policing CSR activity. World business and governing financial body with strong regulatory powers is topical currently as the so-called 'Paradise Papers' where a leak of 13.4 million files from the offshore law firm Appleby is important news. These papers show the worlds' extremely rich, employing accountants to legally avoid paying the tax they owe to the country where they live (Guardian 2017). The disappointing thing in this story is that no criminality is discovered in these cases but instead reveal a state of mind where it appears entirely normal to ignore the broader obligations and responsibilities to society such as those described in this study.

8.2.4 Wind Turbines: The Framework as a Power Tool to Managing Sustainable Development/CSR

It became apparent as the research unfolded that a new coherent pathway or framework was needed in order to guide businesses through the process of managing and evaluating decisions to invest in wind energy projects, and not solely to oversee the project but also to ensure all aspects of SD/CSR were assimilated into the process to ensure that appropriate and responsible business practice was maintained and assured throughout. This unique new framework demonstrated graphically in Fig. 8.3 is used as a coherent pathway for evaluation of CSR/wind turbine investment decision-making. It ensures that robust data is used in corporate decision-making, specifically corporate decisions regarding the assessment of wind turbines and the contribution they make as part of a company's CO_2 emission reduction strategies. Robust empirical wind turbine data encourages critical analysis of wind energy initiatives thereby

Stage:

1. Literature Review of Wind Energy Projects
....

Literature review of wind energy projects through a
corporate social responsibility lens to inform
decision makers of associated risks / benefits
and what to expect in terms of cost / benifit.

2. Decision to Invest in Wind Energy
....

Senior level decision to invest in a wind turbine to
reduce CO2 emissions.
(Triple-Bottom-Line, Elkington, 1997)

3. Data Measurement
....

A technical and economic evaluation and analysis
of wind turbine energy output and wind turbine
power quality output

4. (Wind Energy) CSR/Strategy Integration
....

Identification of key enablers and inhibitors in CSR
and business strategy integration. Information is fed
back and contributes to existing body of knowledge.

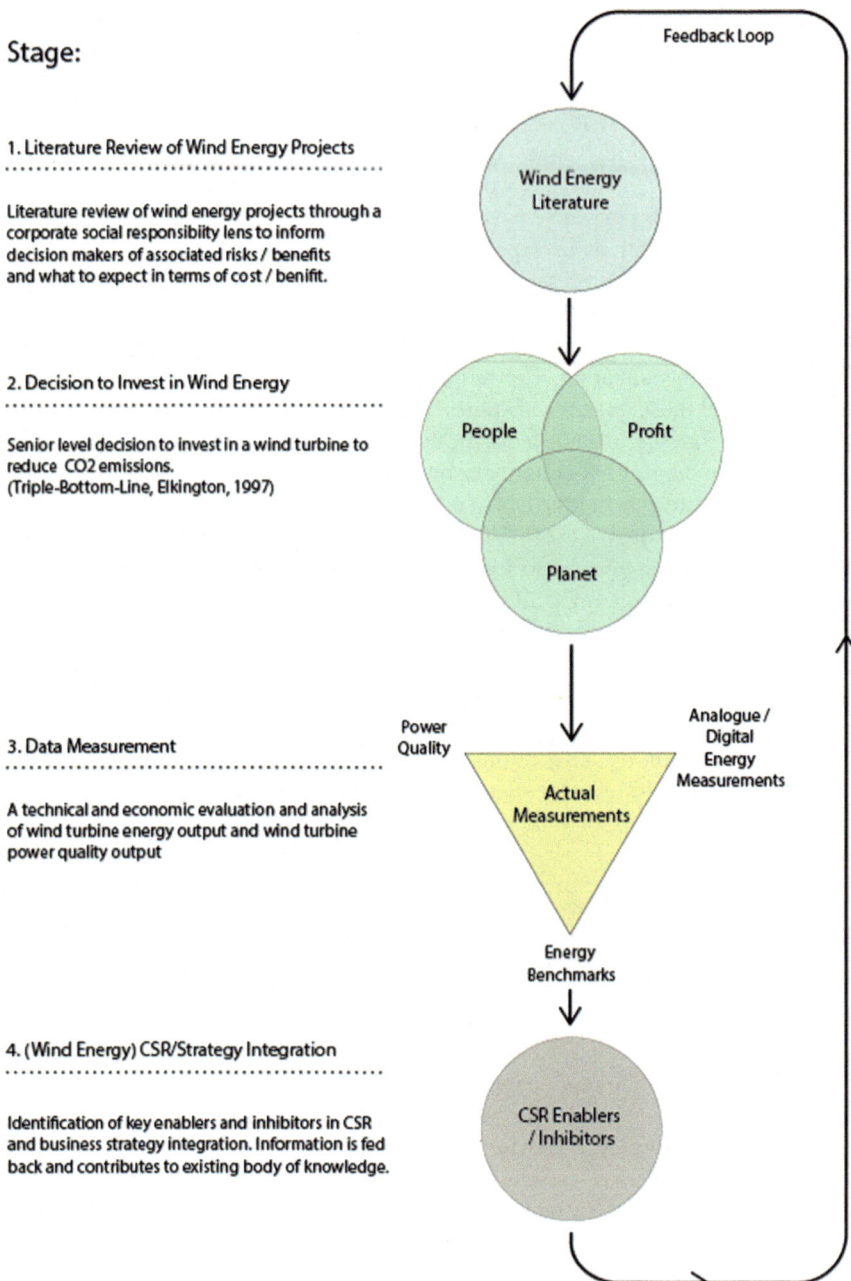

Fig. 8.3 'CSR through the wind turbine lens' evaluation framework

challenging the consensus, and possible shared illusions of wind turbine effectiveness. Such robust data contribute to the critical analysis process by testing reality with actual measured data (Fig. 8.3) as opposed to adopting the poorly informed and poorly empirically supported groupthink consensus regarding perceived benefits of wind turbine technology (Kelman et al. 2017). The evolution of the stages in the development of the new framework is now summarised. Senior-level decisions are made to invest in wind turbine projects (Stage 2, Fig. 8.3) following consultation with a review of the published literature purporting cost and benefit analysis of such projects (Stage 1, Fig. 8.3). Wind turbine decisions are generally made to (i) reduce CO_2 emissions and protect the environment by helping to mitigate climate change and global warming and (ii) reduce energy costs for business operations. Stage 3 (Fig. 8.3) in the evaluation framework is an objective data measurement stage describing the technical and economic evaluation procedures to be carried out that provides actual measured data to contribute to the decision-making evaluation process. The case studies for this data measurement stage (presented in Chaps. 4– 6) are applied to both develop and validate the new framework. Stage 4 (Fig. 8.3) of the framework is where critical enabling and inhibiting factors are identified that allows closer CSR strategy integration with the overall business strategy. In other words, with the information from the literature review and the data measurement stages, there are certain vital factors that contribute to enhancing CSR/strategy integration namely (i) there must be robust, pertinent data from the data measurement stage to contribute to the critical decision-making evaluation process, (ii) there must be a thorough sustainability/CSR knowledge/education of all the cross-disciplinary dimensions of sustainability/CSR.

Overall, by understanding and completing all the individual stages in the framework, wind turbine projects investment decisions are robustly evaluated and managed (Fig. 8.3). The outcome as a result of carrying out each step is fed back into the overall literature section to contribute to the existing body of knowledge and assist current and future potential investors in wind turbine technology as part of their sustainability/CSR efforts to reduce CO_2 emissions and reduce their energy costs.

8.3 Findings Summary Table

Table 8.1 indicates the findings presented through key parameters (left-hand column) as stated in Sect. 8.1 and the associated Chapters where these findings emerge (right-hand column);

Table 8.1 Findings and associated chapters

	Chapter(s)
Wind turbines through a CSR lens	Chapter 2
Wind turbines: a technical and economic evaluation	Chapters 4–6
Wind turbines: the key CSR enablers in integrating CSR and business strategy	Chapters 2, 3, 6, and 7
Wind turbines: the framework as a power tool for managing CSR	Chapters 2, 4–7

8.4 Findings Chapter Summary

This chapter explicitly states the critical findings of the research study. The findings show that sustainability/CSR can be regarded as organisational decision-making that accounts for economic, environmental, and social factors. The technical and evaluation sections showed that fossil-fuel generators appear to be finding it difficult to compensate for the short-term dispersion in parallel-connected wind turbine generators. Better measuring techniques and improved sustainability knowledge can contribute to improved business decision-making. All the studies presented in the book contribute to the development of the novel evaluation framework (Fig. 8.3).

References

Benabou, R. (2013). Groupthink: Collective delusions in organisations and markets. *Review of Economic Studies, 80,* 429–462. https://doi.org/10.1093/restud/rds030.

CSE. (2017). *Sustainability reporting trends in North America 2017.* Centre for Sustainability and Excellence. Available at http://www.sustainability-academy.org. Accessed on 9 December 2017.

Cullen, J. (2013). Measuring the environmental benefits of wind-generated electricity. *American Economic Journal: Economic Policy, 5*(4), 107–133. https://doi.org/10.1257/pol.5.4.107.

Elkington, J. (1997). *Cannibal with forks: The triple bottom line of 21st century business.* Gabriola Island, BC: Capstone.

Guardian. (2017, November 10, Friday). The Guardian view on the Paradise Papers: Not all is lost. *The Guardian.* Available at http://theguardian.com. Accessed on 13 November 2017.

Herve, H. (2018, July). The profession's role in sustainability advocacy challenging CFOs to join the movement. *CPA Journal, 88*(7), 15.

Janis, I. L. (1983). *Groupthink: Psychological studies of policy decisions and fiascos* (2nd ed.). Boston: Houghton Mifflin. ISBN: 0395331897.

Katzenstein, W., & Apt, J. (2009). Air emissions due to wind and solar power. *Environmental Science and Technology, 43*(2), 253–258. https://doi.org/10.1021/es801437t.

Kealy, T. (2014a, April). Financial appraisal of a small scale wind turbine with a case study in Ireland. *Journal of Energy and Power Engineering, 8*(4), 620–627. https://doi.org/10.17265/1934-8975/2014.04.004.

Kealy, T. (2014b, October). Sustainable business development: An Irish perspective. *International Journal of Humanities and Social Science, 4*(12), 166–179.

Kealy, T. (2017, February). Stakeholder outcomes in a wind turbine investment; is the Irish energy policy effective in reducing GHG emissions by promoting small-scale embedded turbines in SME's? *Renewable Energy, 101,* 1157–1168. https://doi.org/10.1016/j.renene.2016.10.007.

Kealy, T., Barrett, M., & Kearney, D. (2015, April). How profitable are wind turbine projects? An empirical analysis of a 3.5 MW wind farm in Ireland. *International Journal on Recent Technologies in Mechanical and Electrical Engineering, 2*(4), 58–63.

Kelman, S., Sanders, R., & Pandit, G. (2017, April). "Tell it like it is": Decision making, groupthink, and decisiveness among US federal subcabinet executives. *Governance: An International Journal of Policy, Administration, and Institutions, 30*(2), 245–261. https://doi.org/10.1111/gove.12200.

Lau, A. K. W., Lee, P. K. C., & Cheng, T. C. E. (2018, July). An empirical taxonomy of corporate social responsibility in China's manufacturing industries. *Journal of Cleaner Production, 188*, 322–338. https://doi.org/10.1016/j.jclepro.2018.04.010.

Melia, P. (2017, November 16). *Ireland is worse country in Europe for taking action to tackle climate change*. News Environment. Available at http://www.independent.ie. Accessed on 18 November 2017.

Chapter 9
Conclusions

Abstract The conclusions of the research study are discussed in this chapter. Sustainable development/CSR is an ambiguous, subjective concept. However, every decision made in the domain of sustainability/CSR must incorporate economic, environmental, and social components. Only decisions that contribute positively to the three individual, yet interlinked, components enable organisations to develop sustainably. The framework, developed and presented in this book, allows decisions to invest in renewable energy technologies, such as wind turbines, to be evaluated robustly. As part of the data measurement stage prescribed in the framework, the power quality measurement aspect exposed a persistently varying, short-term, power output from the generator, regardless of the local wind speed. These short-term variations were calculated in this book as 'coefficient of variation' values. Significant coefficient of variation values reduces the usefulness of the wind turbine electrical generators in offsetting CO_2 emissions associated with embedded fossil-fuel-driven generators. Large power output wind turbines have higher capacity factor values than smaller output wind turbines. The payback periods ranged from twenty-three years for a 10-kW synchronous, three-phase, wind turbine to 6.7 years for a 3.5-MW induction-type (DFIG) wind farm. Measurement of sustainability outcomes and improved sustainability education are crucial factors for intensification in practical sustainability business culture. Future research is recommended in suitable applications for wind turbines whose power output is continuously changing, with some element of smoothing built in to the processes.

9.1 Introduction

This longitudinal ethnographic study took place over eight years. During this time, changes to the global corporate landscape over the short number of years have propelled business activity and business responsibilities into the forefront of the global public narrative. Discussion and debate over such challenges as climate change, extreme weather events, wealth distribution and poverty, scarcity of energy resources, political instability, and food and water shortages are currently high on the public

© Springer Nature Switzerland AG 2020   191
T. Kealy, *Evaluating Sustainable Development and Corporate Social Responsibility Projects*, https://doi.org/10.1007/978-3-030-38673-3_9

agenda. While it may be premature to predict that an economic, social, and environmental perfect storm is brewing, there are unquestionably challenging times ahead (Hillman and Baydoun 2017; Rogers 2017; Shaffer 2017). While it would be unfair and untrue to blame all the worlds' ills on businesses and business activity, multinationals in particular inevitably occupy an influential position as agents of change to make the world a better place for all (Westermann-Behaylo et al. 2015).

Throughout this current study, in-depth analysis and discussion took place describing the social responsibilities of corporations generally referred to as CSR. CSR is the vehicle for achieving sustainable business development. This study highlighted many areas where an opportunity to embrace corporate responsibility within the business sector was evident. In other words, sustainability/CSR should be embraced by the business world as an opportunity to adopt ethical and moral corporate constructs and strive towards what Fremeaux and Michelson (2017) describe as '*conscious capitalism*' and the '*common good*'. One of the practical methods of integrating sustainability/CSR discussed in this study was to make empirical wind turbine data available to other businesses that are considering investing in such similar projects. Dissemination of such data is achieved by the publication of the studies in peer-reviewed journals, books, and conferences. To successfully enable this CSR realisation, businesses must ensure that their empirical information is robust and accurate to assist corporate decisions that contribute to the common good. This book has made essential contributions in providing such data, particularly in the wind turbine/sustainability/CSR area. Grounding a more robust sustainability concept in this study means that businesses that have previously invested in wind turbines possess an opportunity to share the investment experience with other corporations considering investing in similar projects. The data presented in Chapters 4, 5, and 6 evaluating quantity and quality of wind turbine energy/power output fills an identified gap in the literature (Chapter 2), and the businesses who invested in these projects have the opportunity to share their valuable data with other like-minded companies. Grasping these opportunities can enable businesses to contribute to the common good, a principle that involves management taking action that help people achieve their rightful fulfilment. Businesses surely have an ideal opportunity to create *peace through commerce*, as suggested by Greenwood and Freeman (2018), where the authors focused on ethical opportunities for Iraqi businesses.

This chapter lists and discusses the main insights gained as a result of carrying out the study beginning with a reiteration of the research aims and objectives. Successful completion of each of the aims and objectives are discussed in Sect. 9.3. The study makes contributions to knowledge to both the practical and theoretical realms and in doing so has helped to bridge the gap between the business world and the academic world. Future research is subsequently recommended, along with the limitations of this study and a final summary of the work is detailed in Sect. 9.5.

9.2 Reiteration of Research Aim and Objectives

At the outset of this study in Sect. 1.2, the principal research aim and objectives were identified. The aim of the research was to

- develop an analytical closed-loop framework that can be used by businesses to critically evaluate their decision to invest in SD/CSR initiatives specifically concerning wind turbine projects undertaken by the company (**Aim 1**).

The objectives of the research study are to

- review and critique relevant published literature to identify gaps (weaknesses) in the SD/CSR/Wind Turbine space (**Obj. 1**),
- evaluate technical aspects of wind turbine installations (**Obj. 2**),
- carry out economic assessments on wind turbine investments (**Obj. 3**),
- identify key enablers and inhibitors that have the potential to influence the integration of SD/CSR and strategy within an organisation (**Obj. 4**).

Each of the aims mentioned above and the objectives of this book were accomplished during this longitudinal study. Results are presented in Sect. 9.3.

9.3 Accomplishing Research Aim and Objectives

9.3.1 Identify Weaknesses in the CSR Space from the Literature Review

The literature review confirmed the view that sustainability/CSR is an ambiguous, subjective concept. However, there is enough literature published to imply that SD/CSR is a senior management decision-making platform that considers economic, environmental, and social components. Xu et al. (2016) claim that some of the ambiguity problems may lie in the fact that much of the SD/CSR research has little integrity or universality due to the lack of a systematic analytical framework (**Obj. 1**). A new CSR/Wind Turbine systematic analytical framework is presented in this book that demonstrates how wind turbine investment decisions are evaluated. The novel framework introduces an objective measurement stage as a critical component in the development of the framework.

9.3.2 Technical Evaluation of Wind Turbine Projects

The critical insights derived from the technical evaluation in this study are as follows:

- A new closed-loop evaluation framework evolved in this study and is demonstrated and presented in Fig. 8.3. The framework has a specific focus on wind turbines (**Aim 1**).
- Accurate energy output quantity validation using two methods of energy measurement (RDM and Digital) for 10-kW and 300-kW wind turbines presented (**Obj. 2a**). The digital and analogue 10-kW energy measurements matched (Appendices for Chapter 4) but substantial differences were identified in the 300-kW energy measurements (Table 6.7).
- It was found that during low wind speeds, turbines are a 'consumer' of electrical power/energy, importing electricity from the National Grid, see Figs. 6.3, 6.4, and 6.6 (**Obj. 2b**).
- The short-term power output from wind turbines is changing more radically than was previously perceived, on a half-second by half-second basis leading to power quality and ramping power problems, see Figs. 4.3, 6.3, 6.4, 6.5, and 6.6 (**Obj. 2c**). While the wind turbines are certainly producing electrical energy units, the power quality is poor. The power quality parameter that indicates poor quality in this book is the Coefficient Of Variation (COV) parameter. This finding has a knock-on effect on the efficiency of the parallel-connected traditional electrical generators. The traditional back-up generators, much of which are fossil-fuel driven, are unable to respond quickly enough to compensate for the highly dispersant wind turbine power outputs. Therefore, the (mainly) oil and gas usage to drive the back-up generators is not reduced in the expected fashion, reducing the efficiency of the overall system.
- Large output wind turbines have higher capacity factors than small-scale wind turbines, i.e. approximately 30% compared to about 10% (**Obj. 2d**).
- Inductive type wind turbines can cause power factor issues in electrical installations which in turn lead to power quality problems (**Obj. 2e**).

9.3.3 Economic Assessment of Wind Turbine Projects

This study provided some valuable insights into the actual economic benefits gained as a result of deciding to invest in wind turbine technology. The main ideas are listed as follows:

- Connection of an embedded 300 kW asynchronous wind turbine did not reduce the number of imported energy units and hence cost benefits for the SME investor (**Obj. 3a**). This disappointing result is linked to the findings expressed in Obj. 2c where reasons are given for the overall efficiency of wind turbine projects to be less than was expected. The main problem is poor (wind turbine) power quality and the difficulty in compensating for poor quality by traditional (back-up) generators.

- Connection of the 300-kW wind turbine did not reduce GHG emissions for the SME stakeholder and the 10-kW wind turbine installation provided for marginally improved GHG emissions (**Obj. 3b**). The main problem is the reduced power quality generated by the wind turbine, expressed in Coefficient Of Variation (COV) terms.
- An evaluation on a 10-kW three-phase synchronous wind turbine project calculated an IRR of 1.025%, a simple Payback Period of 23 years and a negative Net Present Value for the investment (**Obj. 3c**).
- An economic evaluation of a 3.5-MW wind farm produced a simple Payback Period of 6.7 years for the €5.4 Million investment (**Obj. 3d**). This value is based solely on the quantity of energy units produced. Further research is needed into the quality of the power generated by the wind turbines. While the financial results are satisfactory for the investors, in this case, the national energy benchmarks (g CO_2/kWh) indicate that the substantial increase in wind turbine connected capacity is not reflected in the national (Irish) energy benchmarks.

9.3.4 Key Enablers in Sustainable Development/CSR/Strategy Integration

- Key enabling sustainability/CSR factors such as robust, measurable sustainability data and a strong sustainability/CSR education need to be integrated succinctly and efficiently by business management for sustainability/CSR to penetrate throughout the business (**Obj. 4**).

Table 9.1 is used to summarise the locations (Chapters) in this study where the aim and objectives are accomplished. All the objectives described in Table 9.1 contribute to the aim of this book, i.e. to develop a systematic analytical framework that can be used by businesses to evaluate decisions made to invest in wind turbine technology as part of their GHG emission reduction and cost reduction strategy.

It can be seen from Table 9.1 the interlinked nature of the findings and contributions. The findings/contributions are confirmed in several different chapters which helps to form a coherent document. During this longitudinal study, the researcher used both his technical and business management skills to highlight these valuable research contributions to industry and academia expressed in this chapter.

Table 9.1 Aim and objectives accomplished and associated chapters

Aims and technical objectives accomplished	Associated chapters	Economic objectives accomplished (and Key enabling factors)	Associated chapters
New model: Framework (Aim 1; Obj. 1)	Chap. 1; Chap. 2; Chap. 3; Chap. 4; Chap. 5; Chap. 6; Chap. 7	No reduction in imported energy units with 300 kW turbine (Obj. 3a)	Chap. 4; Chap. 6
Output validation (Obj. 2a)	Chap. 4; Chap. 6	No reduction in GHG emissions (Obj. 3b)	Chap. 4; Chap. 6
Importing electricity (Obj. 2b)	Chap. 4; Chap. 6	Low IRR, long PP, and negative NPV for 10 kW turbine (Obj. 3c)	Chap. 4; Chap. 5; Chap. 6
Rapidly changing power output (Obj. 2c)	Chap. 2; Chap. 4; Chap. 6	Encouraging results for 3.5 MW Wind Farm (Obj. 3d)	Chap. 2; Chap. 5
Importance of turbine size for CF (Obj. 2d)	Chap. 2; Chap. 4; Chap. 5; Chap. 6	Key SD/CSR enabling factors (Obj. 4)	Chap. 1; Chap. 2; Chap. 3; Chap. 4; Chap. 6; Chap. 7
Creation of technical issues (Obj. 2e)	Chap. 4; Chap. 6		

9.4 Linking the Contribution to Knowledge of This Work to the Original Research Aims and Objectives

The aims and objectives were reiterated at the beginning of this chapter (Sect. 9.2). This section highlights how the aims and objectives were achieved through the findings and contributions stated in Table 9.1. Table 9.2 identifies the link between the contributions (Aim 1, Obj. 1, Obj. 2, Obj. 3, Obj. 4, left-hand column) and the research aim and objectives (top row).

A black circle is placed to show how the research aim and objectives on the top row are aligned with the contributions/findings on the left-hand column.

9.5 Future Research

9.5.1 Recommendations

The new strategic sustainable development/CSR, decision-making evaluation framework, presented in this publication utilises wind turbine case studies (10-kW, 300-kW and 3.5-MW) as development and validation tools. However, the model may have replicability to evaluate sustainability development/CSR decisions other than wind

Table 9.2 Link between contributions and research aim and objectives: making the connection

	Identifying gaps in the wind turbine space	Evaluate technical aspects of wind turbine installations	Carry out economic assessments of wind turbine investments	Identification of key enabling SD/CSR factors	Developing a closed-loop framework to manage SD/CSR
New model: Framework (Aim 1; Obj. 1)	•	•	•	•	•
Output validation (Obj. 2a)	•	•			•
Importing electricity (Obj. 2b)		•	•		•
Rapidly changing (Obj. 2c)		•	•		•
Importance of size (Obj. 2d)		•			•
Creation of technical issues (Obj. 2e)		•			•
No reduction in energy units (Obj. 3a)		•	•		•
No reduction in GHG emissions (Obj. 3b)	•	•		•	•
Poor economic results 10 kW turbine (Obj. 3c)	•		•	•	•
Encouraging results for 3.5 MW wind farm (Obj. 3d)			•	•	•
Key SD/CSR enabling factors (Obj. 4)	•			•	•

turbine investment decisions made by business managers (Kealy 2020). The three main stages to follow, in all TBL evaluations, are (i) review the literature on the specific area to ascertain the potential risks and benefits associated with the investment decision, (ii) after, and if, the decision is made to implement the initiative, carry out an objective measurement on the variable(s) by which the decision is evaluated, (iii) identify critical issues in the organisation which have the potential to empower the integration of the sustainability strategy with the strategy of the organisation. For example, if decisions are to be evaluated in the 'Biodiversity' strain, then the following stages are to be followed; Environment—(Literature on) Biodiversity—(Objective measuring stage), e.g. Measure diversity of species in an area—Identify key CSR enablers within the organisation. As with the case of the wind turbine projects examined for this research study, the results of the output from the numbers of stages in the Biodiversity strain provide feedback to senior decision-makers to assist and evaluate their Biodiversity decision-making process.

The variations in the turbine power output data presented in Fig. 4.3 and Figs. 6.3, 6.4, 6.5, and 6.6 point to a necessity for future research in the area of energy smoothing. While energy storage is currently receiving significant interest in both academia and practice, energy (short-term) smoothing research is not as proactive. It is possible that energy smoothing applied to the 10-kW and 300-kW wind turbine installations would increase the overall efficiency of the electrical system. The possibility of a link between significant variations in turbine output power and the mostly unchanging energy loss benchmarks for electricity generation in Ireland (Table 1.2) should be investigated. The considerable changes in the turbine power output rendered it problematic for the parallel-connected fossil-fuel electrical generators to accommodate the variations. However, while this current book focuses on wind turbines used as embedded electrical generators, there is scope for more empirical research to be carried out into the potential for stand-alone (not embedded with the national electricity grid) wind turbines to provide the energy source to electrolysers used to produce low-carbon hydrogen. As discussed in Sect. 1.1.2, local, renewable-based low-carbon hydrogen can be used to heat homes, and in the iron, steel, transport, and chemical industries. This future research can ascertain if, and to what extent, the short-term variations in the turbine power output (Fig. 4.3 and Figs. 6.3, 6.4, 6.5, and 6.6) affect the overall electrolysis process.

Future research is also recommended on the quality and quantity of CSR/Sustainable Development modules that are currently taught on programmes in academic institutions. This research could focus on the ratio between financial modules and non-financial modules on undergraduate programmes.

9.5.2 Limitation of This Study

A limitation with this study is that a complete 'Life Cycle Assessment' (LCA) was not undertaken for the selected wind turbine projects, with the operational stage the only stage of its life cycle being evaluated. It is recommended that future research is carried out considering an LCA on wind turbines, i.e. where they are purchased, how it was manufactured, and the raw materials used and finally how it is disposed of/recycled at its end-of-life stage.

This study is limited by the fact that most of the participants in this research were associated with established sustainability reporting databases, specifically the Global Reporting Initiative (Global) and Origin Green (Ireland). It did not include those companies who chose not to report at all on their sustainability efforts, for whatever reasons.

Ireland is a small island nation with a small population, it is possible that the 'groupthink' culture discussed throughout the book (Heaslip et al. 2016; Waters 2015; Sims and Sauser 2013) may have possibly expedited an unchallenged acceptance of wind turbines as the main answer to the problem of CO_2 emissions. This phenomenon may not be replicated in other cultures or countries.

9.5.3 Final Summary

This sustainable development/CSR ethnographic research study encompasses several academic disciplines in the research process, namely the disciplines of Management, Engineering, and Economics. The interdisciplinary nature was identified as a critical component in the efforts to propagate sustainability/CSR activity within organisations. From the literature, sustainable development/CSR is deemed to be a business decision-making platform that incorporates economic, environmental, and human facets. The three interdependent dimensions have been modelled by Elkington (1997) as the Triple-Bottom-Line (TBL) accounting framework. The novel closed-loop evaluation framework presented in this book is a development of the, sometimes criticised, TBL model proposed by Elkington (1997). The framework is focused on one of the TBL criticisms, namely the measurement aspect, and includes an objective measuring part that contributes significantly to the decision evaluation process. During the *data measurement* stage (Power Quality measurement in Fig. 8.3), significant short-term variations were identified in the power output signals from the wind turbines. The variations were expressed as statistical coefficient of variation measures. It was found that parallel-connected fossil-fuel electricity generators were having difficulty responding fast enough to compensate for the short-term variations from the wind turbines. This finding meant that the efficiency of the overall electrical system did not improve significantly despite the (significant) increase in the number of wind turbines connected to the National Grid.

Wind turbine power output incorporating smoothing (capacitors) has the potential to make far more significant contributions to the CO_2 reduction efforts that those currently in service. The findings and results of the evaluation process for the wind turbine projects utilised in the validation process, particularly the 10-kW and 300-kW turbines, were disappointing from an economic and environmental aspect, and therefore by definition did not create benefit for the societal component of sustainability. However, the turbines could potentially contribute to providing power to heating loads and loads that could use local, renewable-based hydrogen such as those found in the iron, steel, and chemical industries. In these specific loads, short-term dispersion may not be as significant a problem as it is with embedded fossil-fuel-driven electrical generators and their efforts to compensate for the rapidly changing turbine output signals.

The assortment of data sources utilised in this ethnographic study ranging from literature reviews, case studies, surveys, and interviews enabled observation of the uncertainty encompassing the TBL concept. This uncertainty was manifest regarding the real benefits linked with companies embracing embedded wind turbine projects. This study found a 'groupthink' culture and a shortage of wisdom exists in the discussion on sustainable development. A more independent critical analysis is needed into the practicalities surrounding wind turbine projects. It may be wise to listen to a broad range of non-traditional business stakeholders in the sustainable development discussion as the area has such a broad implication, from economic, environmental, and social backdrops.

References

Elkington, J. (1997). *Cannibal with forks: The triple bottom line of 21st century business.* Gabriola Island, BC: Capstone.

Fremeaux, S., & Michelson, G. (2017). The common good of the firm and humanistic management: Conscious capitalism and economy of communion. *Journal of Business Ethics, Issue, 145,* 701–709. https://doi.org/10.1007/s10551-016-3118-6.

Greenwood, M., & Freeman, R. E. (2018). Deepening ethical analysis in business ethics. *Journal of Business Ethics, 147*(1), 1–4. https://doi.org/10.1007/s10551-017-3766-1.

Heaslip, E., Costello, G. J., & Lohan, J. (2016). Assessing good-practice frameworks for the development of sustainable energy communities in Europe: Lessons from Denmark and Ireland. *Journal of Sustainable Development of Energy, Water and Environmental Systems, 4*(3), 307–319. https://doi.org/10.13044/j.sdewes.2016.04.0024.

Hillman, J. R. & Baydoun, E. (2017). Food security in an insecure future. In S. Murad, E. Baydoun, & N. Daghir, (Eds.), *Water, energy & food sustainability in the Middle East* (pp. 261–282). Cham: Springer. https://doi.org/10.1007/978-3-319-48920-9_12.

Kealy, T. (2020). A closed-loop renewable energy evaluation framework. *Journal of Cleaner Production, 251,* April. https://doi.org/10.1016/j.jclepro.2019.119663.

Rogers, P. (2017). The triangle: Energy, water & food nexus for sustainable security in the Arab Middle East. In S. Murad, E. Baydoun, & N. Daghir (Eds.), *water, energy & food sustainability in the Middle East* (pp. 21–43). Cham: Springer. https://doi.org/10.1007/978-3-319-48920-9_2.

Shaffer, L. J. (2017). An anthropological perspective on the climate change and violence relationship. *Current Climate Change Reports, 3*(4), December, pp. 222–232. https://doi.org/10.1007/s40641-017-0076-8.

Sims, R. R., & Sauser, W. I. (2013). Towards a better understanding of the relationships among received wisdom. *Groupthink, and Organisational Ethical Culture, Journal of Management Policy and Practice, 14*(4), 75–90.

Waters, J. (2015, 5th April). Nuance or subtlety unwelcome at an RTE that's still in groupthink's grip. *Sunday Independent*, News, p. 23.

Westermann-Behaylo, M. K., Rehbein, K. and Fort, T. (2015). Enhancing the concept of corporate diplomacy: Encompassing political corporate social responsibility, international relations, and peace through commerce. *Academy of Management Perspectives*, 29(4), November, pp 387–404. https://doi.org/10.5465/amp.2013.0133.

Xu, J., Li, L. & Zheng, B. (2016). Wind energy generation technological paradigm diffusion. *Renewable and Sustainable Energy Reviews*, 59(June), 436–449.

Conferences at Which Author Presented a Paper

Kealy, T. British Academy of Management (2019) 'Corporate Sustainability Reporting: The Practical Implications,' Aston University, Birmingham, UK, 3–5 September 2019.

Kealy, T. International Sustainable Ecological Engineering Design for Society (SEEDS) Conference (2018) 'Key Enablers in the CSR/Business Strategy Integration Space', Dublin Institute of Technology, 6–7 September 2018. Published in 'International Sustainable Ecological Engineering Design for Society [SEEDS] Conference 2018', by LSI Publishing, for Lloyd Scott and Chris Gorse, September 2018, ISBN Number: 978-0-9955690-3-4.

Kealy, T. Irish Academy of Management (IAM) Annual Conference (2014), 'Sustainable Business Development: an Irish Perspective' University of Limerick, 3–5 September 2014.

Kealy, T. Irish Academy of Management (IAM) Annual Conference (2013), 'Do Middle Managers Contribute to their Organisations Strategy?' Waterford Institute of Technology, 2–4 September 2013.

Kealy, T. 48th International Universities Power Engineering Conference (UPEC) (2013), 'Small-Scale Wind Turbines: An Appraisal', Dublin Institute of Technology, September 2nd to September 5th, pp. 1–6, https://doi.org/10.1109/upec.2013.67.15004.

© Springer Nature Switzerland AG 2020   203
T. Kealy, *Evaluating Sustainable Development and Corporate Social Responsibility Projects*, https://doi.org/10.1007/978-3-030-38673-3

Appendices for Chapter 3

List of questions used in this research:

Question 1: Milton Freeman famously said that businesses have one, and only one, social responsibility, namely to increase or generate profit. Would you agree or disagree with this statement?

Question 2: Does your organisation/business view sustainability as an important aspect of the running/management of the organisation?

Question 3: Is sustainability seen as a strategic issue within your organisation/business?

Question 4: What do you see as the top global sustainability challenges?

Question 5: Were there any external factors that caused your organisation/business to look at sustainability issues?

Question 6: How would you evaluate/measure the success of your sustainability initiatives?

Question 7: Is your organisation/business aligned with any accrediting body in the sustainability area?

Question 8: What methods or processes have been used to progress sustainability issues within the organisation/business?

Question 9: Do you think that ethical business practices contribute to the sustainability efforts of your organisation?

Question 10: How do you view the concept of 'Social Entrepreneurship'?

Question 11: Do you see the personal development of each employee as part of your sustainability efforts?

© Springer Nature Switzerland AG 2020   205
T. Kealy, *Evaluating Sustainable Development and Corporate Social Responsibility Projects*, https://doi.org/10.1007/978-3-030-38673-3

Codes Generated for Chapter 3

VPCS = Varied Perceptions of the Concept of Sustainability

Code	Theme	Sub-theme
More equal society	VPCS	Social equality
Good quality of life	VPCS	Social equality
Charitable mission	VPCS	Social equality
Light Emitting Diodes (LED's)	VPCS	Energy
Building Management System (BMS)	VPCS	Energy
More efficient logistics operation'	VPCS	Energy
Feed the seven billion people	VPCS	Food
Not hungry	VPCS	Food
Food and the production of food is a challenge	VPCS	Food
Ethical	VPCS	Ethics/morals
An ethical business is a moral business	VPCS	Ethics/morals
Moral awareness of sustainability	VPCS	Ethics/morals
Energy efficiency policy	Regulation	
Government must be very involved	Regulation	
Reduce CO_2 emissions in line with the UK directive	Regulation	
Government policy'	Regulation	
Legislation	Regulation	
Change in legislation	Regulation	
Flag-ship of the organisation	Leadership	
Leadership is very important	Leadership	
Good leadership	Leadership	
Leadership is very important	Leadership	
Leader	Leadership	
(PLC), investors	Marketing	
Marketing of the business	Marketing	
People are what make us	Human aspects	
We also treat each employee with respect	Human aspects	
Human beings	Human aspects	
Personal development	Human aspects	
People must be treated with respect	Human aspects	
Important aspect is the people aspect	Human aspects	
Local people have taken a chance	Human aspects	
It encourages camaraderie	Human aspects	

33 codes in total were generated.

Appendices for Chapter 4

Testing the new closed-loop model; the actual measurements section of the loop shows 'Digital Energy Measurements' and 'Analogue Energy Measurements' (in addition to 'Energy Benchmarks' and 'Power Quality'). This triangulation of the data provides robustness to the closed-loop. The digital energy outputs are taken from the two inverters shown in Fig. A4.1. The analogue energy measurement is taken from the RDM shown in Figs. A4.2 and A4.3.

Comparison of Digital energy output indicators with the Rotating Disc Meter (Electromechanical, analogue) energy output indicator;

Between Saturday 9 April 2016 and Wednesday 18 May 2016, the difference in the Rotating Disc Meter (RDM) is 17.9 ($\times$40 CT Ratio) = 716 kWhs.

The difference in Left-Hand-Inverter digital indicator is 434 kWh

The difference in Right-Hand-Inverter digital indicator is 305 kWh

Total digital energy output indicators are 739 kWh [(434 + 305) kWh]

RDM shows (716/736) = **97% accuracy**. Both the digital and analogue energy measurements are valid in this case.

Fig. A4.1 Inverters with digital indicators on the front panel (Kealy)

T. Kealy, *Evaluating Sustainable Development and Corporate Social Responsibility Projects*, https://doi.org/10.1007/978-3-030-38673-3

Fig. A4.2 Rotating Disc
Meter (RDM) validating
digital energy readings
(Kealy)

Energy Benchmarks (Imported kWh energy units per annum)
   2009—77,312-kWh; 2010—77,064-kWh; 2011—68,519-kWh
   2012—76,338-kWh imported with the turbine connected. Turbine produced 7260-kWh units. Percentage improvement is (76338)/(83598) = 9% in 2012 (Table A4.1).

SEAI Method of Calculating Annual Carbon Intensity Value
To calculate the annual carbon intensity value, SEAI first surveys all the electricity generators for their fuel usages and electricity produced figures, which SEAI check against emissions trading figures to verify. SEAI then take those fuel quantities and use emission factors which they get directly from the EPA to calculate the amount

Table A4.1 Carbon intensity of electricity benchmarks in Ireland (SEAI)

Year	g CO_2/kWh
2017	437
2016	482.6 (483)
2015	467.5 (468)
2014	456.6 (457)
2013	466
2012	529
2011	489
2010	530
2009	522
2008	547
2007	560
2006	596

Fig. A4.3 Rotating Disc Meter used for analogue energy measurement (Kealy)

Table A4.2 NPV for 4% Interest Rate for 10 kW Wind Turbine Project

Time	Cash flow (€)	Discount factor (4%)	Present value (€)
Immediately	−26,620	1	−26,620
1 years' time	1142	0.96	1096
2 years' time	1142	0.925	1056
3 years' time	1142	0.89	1016
4 years' time	1142	0.855	976
5 years' time	1142	0.82	€936
6 years' time	1142	0.79	902
7 years' time	1142	0.76	799
8 years' time	1142	0.73	833
9 years' time	1142	0.703	802
10 years' time	1142	0.675	770
11 years' time	1142	0.649	741
12 years' time	1142	0.625	713
13 years' time	1142	0.6	685
14 years' time	1142	0.577	658
15 years' time	1142	0.555	633
16 years' time	1142	0.533	608

(continued)

of CO_2 produced, and they then divide that by the amount of electricity produced to get carbon intensity (SEAI 2018) (Table A4.2).

Table A4.2 (continued)

Time	Cash flow (€)	Discount factor (4%)	Present value (€)
17 years' time	1142	0.51	582
18 years' time	1142	0.49	559
19 years' time	1142	0.474	541
20 years' time	1142	0.456	520
21 years' time	1142	0.438	500
22 years' time	1142	0.42	479
23 years' time	1142	0.405	462
24 years' time	1142	0.39	445
25 years' time	1142	0.375	428
	2000	0.375	750
		NPV	**−9380**

Appendices for Chapter 6

[Total of three sealed energy kWh meter readings between 12 February 2016 and 16 March 2016 is 7068.6 kWh units and RDM under test reads 6924 kWh units for the same period. This gives an overall accuracy for the RDM of 98%] (Figs. A6.1, A6.2, and A6.3).

Energy Benchmarks for 2016

Tonnage Output 80,568 Tonnes
Electrical Energy Imported 1,644,252 kWh
kWh/Tonne Benchmark 20.41 kWh/Tonne

Energy Benchmarks for 2017

Turbine produced 276,433 kWh electrical energy units in 2017.
Electrical Energy Imported 1,545,600 kWh from the national grid in 2017 (1,079,000 Day units and 466,572 Night units)
Total Exported units were 11,532 kWh units (1464 Weekday units and 10,068 Night and Weekend units) (Fig. A6.4).

Wind Turbine and Imported Energy Values for 2018

Wind Turbine produced 346,698 kWh units of electrical energy in 2018
Day Imported Units = 1,170,180 kWh, Night Imported Units = 509,028 kWh units;
Total Imported kWh energy units for 2018 = 1,679,208 kWh units (Original factory).
Tonnage output in 2018 was 107,191 tonnes, this is the two adjoining facilities.
MPL said they imported 1,903,908 kWh units in 2018 (two facilities) (Tables A6.1, A6.2, and A6.3).

© Springer Nature Switzerland AG 2020
T. Kealy, *Evaluating Sustainable Development and Corporate Social Responsibility Projects*, https://doi.org/10.1007/978-3-030-38673-3

Fig. A6.1 Validating data
from Rotating Disc Meter
(RDM) with sealed
calibrated utility energy
meters (Kealy)

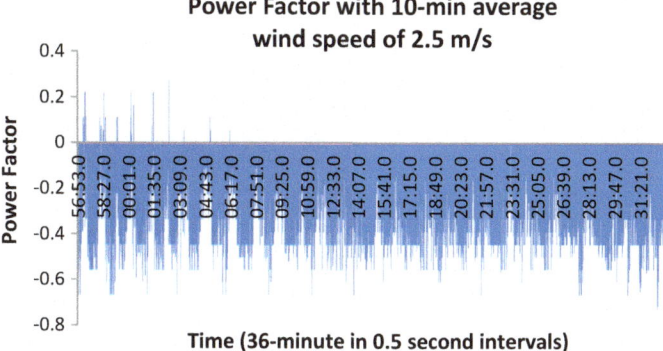

Fig. A6.2 Turbine Power Factor (PF) on Thursday 5 November 2015 with a 10-min average wind speed of 2.5 m/s (negative PF values indicate turbine acting mainly as an electrical load under these low wind conditions)

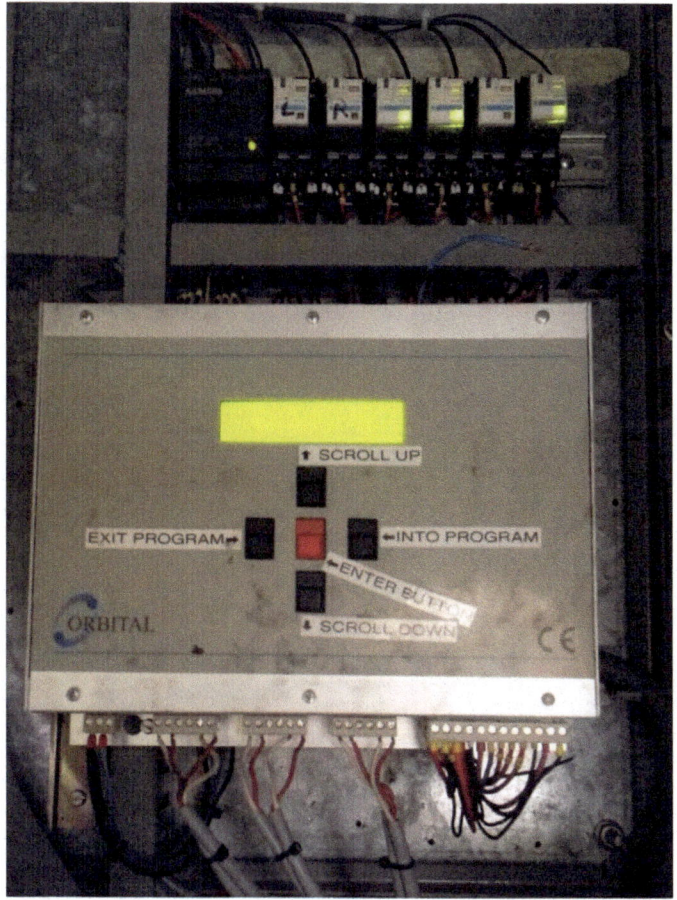

Fig. A6.3 Three-phase digital energy meter (DM) in main panel at base of turbine tower (Kealy)

Fig. A6.4 Three-Phase
Rotating Disc (Analogue)
Energy Meter (RDM) in
EGIP relay panel in main
electrical switch-room
(Kealy)

Table A6.1 Wind speeds—30th May to 10th June for Dublin Airport/Dunsany/Mullingar

2014	Wind speed in m/s		
	Dublin Airport	Dunsany	Mullingar
30th May	3.7	2.4	1.5
31st May	3.4	2.6	1.9
1st June	4.3	3.4	3
2nd June	4.1	3.3	2.2
3rd June	4.1	3.3	2.6
4th June	5.6	3.9	2.8
5th June	4.3	2.6	1.9
6th June	5.2	4.5	4.1
7th June	4.6	3.4	3.7
8th June	5.9	4.4	4.4
9th June	5.8	3.3	3.7
10th June	3.7	3.2	2.7
Average (m/s)	4.6	3.4	2.9

Table A6.2 Wind Speeds—13th June to 24th June for Dublin Airport/Dunsany/Mullingar

2014	Wind speed in m/s		
	Dublin Airport	Dunsany	Mullingar
13th June	2.8	2	1.5
14th June	3	2	1.7
15th June	3.7	2.5	1.7
16th June	3.5	2.7	2.2
17th June	2.3	1.8	1.5
18th June	2.7	2.3	1.9
19th June	4.8	3	2.3
20th June	3.3	2.3	1.9
21st June	3.7	2.5	2.1
22nd June	3.5	2.4	1.8
23rd June	2.8	1.9	1.7
24th June	2.3	1.7	1.5
Average (m/s)	3.2	2.3	1.8

Table A6.3 Monthly average wind speeds for 2014 and May/June 2013

	DA	DA	Dunsany	Dunsany	Mullingar	Mullingar
	WSp Knots	WS m/s	WSp Knots	WS m/s	WSp Knots	WS m/s
January 2014	12.7	6.5	9.8	5	7.3	3.8
February 2014	15.9	8.2	12.3	6.3	9.3	4.8
March 2014	12.1	6.2	9.1	4.7	6.6	3.4
April 2014	9.8	5	8	4.1	6.3	3.2
May 2014	9.9	5.1	7.5	3.9	5.6	2.9
June 2014	7.6	3.9	5.5	2.8	4.6	2.4
July 2014	8.7	4.5	6.5	3.3	4.9	2.5
August 2014	11.4	5.9	8	4.1	5.7	2.9
September 2014	6.5	3.3	5.1	2.6	3.9	2
October 2014	11.5	5.9	8.6	4.4	6.5	3.3
November 2014	9.1	4.7	6.5	3.3	5.2	2.7
December 2014	13.9	7.2	9.7	5	6.8	3.5
		5.5		4.1		3.1
May 2013		6.1		4.7		3.6
June 2013		4.8		3.7		2.8

Appendices for Chapter 7

Appendix 7A

17-Questions in Survey

Q1. In which of the following industries do you operate?

- Agriculture, Forestry, Fishing
- Mining, Quarrying, and Oil and Gas Extraction
- Utilities
- Construction
- Manufacturing
- Wholesale
- Retail
- Transportation and Warehousing
- Finance and Insurance
- Real Estate and Renting and Leasing
- Waste Management
- Educational Services
- Health Care
- Public Administration
- Other (please specify)

Q2. How many employees does your organisation have?

- Between 1 and 9
- Between 10 and 49
- Between 50 and 249
- Between 250 and 999
- Between 1000 and 9999
- Between 10,000 and 100,000
- Greater than 100,000

© Springer Nature Switzerland AG 2020
T. Kealy, *Evaluating Sustainable Development and Corporate Social Responsibility Projects*, https://doi.org/10.1007/978-3-030-38673-3

Q3. As the CSR contact in your organisation are you at 'middle manager' or 'top manager' (strategic) level?

- Middle manager
- Top manager
- Other level (please specify)

Q4. How many years of experience do you have working in the CSR/Sustainable Development area?

- Less than 1 year
- Between 1 year and 5 years
- Between 5 years and 10 years
- Greater than 10 years
- Other relevant experience (please specify)

Q5. Which of the following levels of engagement best describes your Corporate Social Responsibility/Sustainable Development activities within your organisation?

- Upper level: Engage in responsible actions for the common good as explicit part of your mission
- Middle level: report voluntarily and strive for sustainability rankings
- Lower level: doing the minimum to comply with regulations
- None of the above

Q6. State the sustainability reporting framework to which your company is aligned, if any?

Q7. Comment on the advantages, and disadvantages, of sustainability reporting in your organisation. Feel free to make recommendations on suggested improvements that could be made to this aspect of CSR/Sustainable Development:

Q8. Identify your highest personal academic business qualification from the following list:

- Certificate
- Diploma
- Degree
- Masters
- PhD

Q9. Where did you complete your business academic training?

- United Kingdom
- Ireland
- Europe (excluding UK and Ireland)
- United States of America
- Other country (please specify)

Q10. When did you complete your most recent academic training?

- Within the last 2 years
- Between 2 years and 5 years ago
- Between 5 years and 10 years ago
- More than 10 years ago
- Never completed formal training

Q11. As part of your academic training, did you complete modules on 'Leadership'?

- Yes
- No

Q12. As part of your academic training, did you complete modules on 'Corporate Social Responsibility/Sustainable Development'?

- Yes
- No

Q13. Do you think that religions and faith-based organisations, for example Religious Social Teaching, can have a positive influence on your business development?

- Yes
- No

Q14. Comment on the level of support you receive in your CSR/Sustainable Development duties within your organisation.

Q15. In your view, should governments have more influence in promoting sustainable business development?

- Yes
- No
- Feel free to comment on this issue

Q16. In your view, what are the major barriers preventing Corporate Social Responsibility/Sustainable Business Development managers from leading their businesses up to a higher sustainability level as suggested in question 5?

Q17. If you wish, feel free to add your comments on any pertinent issues arising from, or in addition to, the questions in this survey:

Appendix 7B

Appendix 7B (advantages/disadvantages)

	Advantages	Disadvantages
1	Allows us to remain connected to our stakeholder groups [SE], provides ongoing data for which continuous improvement [IM] plans can be designed and measured, communicates to external audiences that our organisation takes sustainability seriously [R]	Some may feel that reporting in itself 'is sustainability', much energy and resources dedicated to reporting can siphon from actual action plans [SF], G4 standards [NS] are very loose—companies can hide behind choosing to not report certain metrics

(continued)

(continued)

	Advantages	Disadvantages
2	Provides a comprehensive picture that shows the interconnected nature of sustainable business objectives [S]	Extreme amount of resources [RST] required to address many criteria of less importance to stakeholders or not as applicable to U.S. issues
3	It shows a well-managed and transparent company [R] and can provide good feedback on the gaps that may exist [IM]	It is time-consuming [RST] and there are more and more raters and rankers [NS] popping up all the time with often overlapping data requests
4	Drives innovation [IM]; transparency; builds credibility, trust; enhanced reputation [R]; brand; mitigates risk; shareholder activism [SE]; supports business strategy [S]; employee retention/recruitment [H]	Demand for disclosure grows
5	Alignment with corporate mission [S], engagement with stakeholders [SE], provides information to investors [SE]	Cost and complexity[CX], and a lack of common standards [NS] within and between industries
6	The opportunity to share with stakeholders the positive contributions we are making to society. It's also an opportunity to demonstrate the challenges we face, and how we are managing these and the risks associated with them [SE]	There are few disadvantages to reporting. However, there are considerable resources [RST] that go into the process
7		The main challenge is to adapt the frameworks like GRI G4 to the company [NS] of our size (small to medium), we don't always have all the information available [CX]
8	Data is available to influence decision-making [TM] but needs to be business team relevant to really achieve this. Corporate data often does not mean anything to anyone internally in particular [SF]. Data also needs to highlight a trend that is either advantageous or disadvantageous to the business or be something that either affects the financial bottom line, reputation [R], or long-term value	Very time-consuming [RST] and a perceived burden on some areas of the business with no real value other than ticking some reporting boxes [SF]. (See recommendation over)
9	It's the right thing to do [E]. Full audience engagement [SE]—internal [H] and external. Shows our commitment to values and ethical [E] business. Engages customers [SE] and provokes innovative thinking [IM]	Collating the information. Measuring the impact of the report [M]

(continued)

(continued)

	Advantages	Disadvantages
10	Enhanced reputation [R]	We report on some aspects of our sustainability credentials, however, do not currently have the impetus or resource [RST] to report on this externally
11	People are becoming more readily aware of sustainability in business. Clients are frequently asking me about it and it is good that we include this in our MI (Management Information) reviews [IM] with our Clients [SE] so that we can share with them	The frameworks do not always allow for a focus on material issues so you end up producing many reports for the different audiences [NS]. Alignment/flexibility would help
12	Used a communications tool to differentiate the business, which generates sales and better working relationships with customers [R]	As a very small organisation, it is quite time-consuming and resource [RST] intensive. Finding more efficient ways throughout the year to track and monitor our impact would make it less demanding
13	Ensures we maintain focus on CSR both internally [H] and externally. Meets customer and shareholder requirements[R][SE]	Resource/cost [RSF]
14	Excellent way to ensure we focus on the material issues relating to sustainability [S], with proactive management of each, including targets and ambitious, and continuous improvement [IM] programmes—way of communicating with our stakeholders [SE] about our values, strategy [S], performance—demonstrates best practice integrity [E]—i.e. in line with the advice we give clients	Time [RST] and cost [RSF] of having people in the business to do this
15	Sustainability reporting is positive as it drives change [IM] in the company, engages stakeholders [SE], and helps us to demonstrate our key areas of focus	Reporting is often driven by indices, e.g. DJSI, FTSE4Good, EIRIS, etc. [NS} so disclosure can often be more complex [CX] and less accessible to all stakeholders
16	As a sustainable waste management company CSR reporting is critical to our business model [S]	
17	Internally [H], sustainability reporting has allowed us to monitor the performance of our business. Externally, the real advantage is that we can demonstrate our progress (as well as challenges) to external stakeholders [SE]	

(continued)

(continued)

	Advantages	Disadvantages
18	There is good intent throughout the organisation, and we have a strong ethos [E] of CR throughout the staff [H] with high levels of participation in fundraising/volunteering, etc., which includes rewarding staff for their efforts. We struggle a little to implement initiatives, particularly regarding environmental sustainability, which would change typical staff behaviour, processes, or procedures which is to be expected	
19	Example of leadership, maintain CSR on the agenda of top management [TM] and enable a better understanding of CSR across the organisation	No real disadvantage other than the time [RST] it takes
20	Makes all aware of our targets [IM] and responsibilities	
21	Alignment with business development strategy [S] and Ethical [E], Economic, and Environmental activities	Accountants usually don't see the benefits [M]
22	Many customers now requesting details of our sustainability programme [SE], by having the programme up and running, we have the necessary information they require	
23	Sustainability reporting is very important as the primary platform from which all other functions communicate with our staff [H]. Staff involvement is fundamental for us in achieving our sustainability reduction targets [IM] as part of our Origin Green plan	
24	Written reports help with training [H], uniformity, and proof of efforts towards sustainability [R]	Written reports can be negative towards newer innovative ideas from staff who often may not have the confidence to challenge a written process. Record keeping can be difficult [CX] to do with sustainability in mind
25	Origin Green is a great structure by which to track and report our sustainability. While the company has always been environmentally responsible, Origin Green gives us a structure which demands targets and compliance [R], which is good for us to further develop our commitment [E]	There are no disadvantages to this structure that we can see

(continued)

(continued)

	Advantages	Disadvantages
26	There is a lot of historical evidence of sustainability initiatives to provide good case studies [IM]	However, the lack of a structured approach to data gathering [NS] in line with a sustainability framework means it is more challenging [CX] to develop data-driven reporting. This is currently being developed
27	Creates a positive energy and feel-good factor [H] can be used in marketing [M]	Can lead to satisfaction with achievements rather than striving for more difficult goals [SF]
28	We are just 6 months into the Origin Green programme so currently, we are learning how the reporting framework functions; however, it seems relatively straightforward	
29	Our customers want it [SE]. Improves morale amongst employees [H]. Correct to carry out sustainability activities. Reduces cost. Improves competiveness [IM]	
30	Good PR [M], Promotes and encourages an ethical [E] ethos throughout the company. Fosters good corporate governance [R]	Very resource [RSH] and time [RST] intensive, especially for small companies
31	Delivers value to our customers/positive contribution to the local community [SE]	Can't think of any
32	Making periodic reports pushes us to evaluate and measure in a structured way [IM]	As time [RST] is such a limited resource in a small company, it is difficult to give this type of activity priority
33	It formalises that which we have been doing for the past 30 years [S]. Any company can get involved, but I do believe the social bodies/charities need to up the ante and provide packages in line with the CER requirements of companies, if both are to develop and move forward [SE]	
34	More of our customers are looking for products produced responsibility [SE] and sustainable so we can market this [M]. It reduces cost	It can be time-consuming [RST] when working in a small organisation as we do not have the staff [RSH] to dedicate someone to work full time on the role
35	Improve employee morale [H]. Improve community sentiment [M]. Improve organisational culture [E]	
36	Enhances corporate image [M], aligns with the strategies [S] of our customers. Creates a positive culture/environment for staff [H]. Yields financial return. Publishing/reporting can lead to challenges from competitors and NGOs around the areas the organisation is performing less well on [IM]	

(continued)

(continued)

	Advantages	Disadvantages
37	By having to report, we have to keep track of our usages by recording [IM]	
38		It is very tedious and takes a long time [RST]. Some of the categories which we report to are very specific, and others are sometimes not financially viable
39	Shows you where you can save money and improve [IM]	It takes a lot of time [RST]
40	Awareness and information available to make conscious environment decisions [S], such as reducing our carbon footprint. Enhancing the quality of conditions for our employees [H] with a greener environment. Determining the costs and delivering efficiencies [IM]. Sustainability plan in place with targets. Employees are assets of our business and worker satisfaction is very important, we have cycle-to-work schemes, medicals provided every 2 years and have the health and welfare well-being of our employees as a priority [H]. Regular training and records	
41	It helps with B2B sales [M]. Can make the business more efficient [IM]	Consume time [RST] and Money [RSF] at times
42	Savings on waste/water/energy costs [IM]. Staff awareness is improved [H]	Time-consuming [RST] reporting requirements as we do not have someone [RSH] solely hired for environmental/CSR/Sustainable Development
43	Advantages to include employee engagement [H], strategic planning [S], capital investment	

Codes:

SE—Stakeholder Engagement (external)
IM—Improvement
R—Reputation
H—Human (internal)
E—Ethics
S—Strategy
TM—Top Management
M—Marketing
NS— Non-Standard Reporting
SF—Surface Efforts
RST—Resources (Time)

RSH—Resources (Human)
RSF—Resources (Financial)
CX—Complexity
M—Measurement

Appendix 7C

Appendix 7C: Responses to faith input to business decisions.

1. Our global organisation does business in over 100 countries. It would be difficult to work with faith-based organisations that are non-affiliated and inclusive enough to meet the needs of our diverse stakeholders [RK].
2. I'm not sure that the answer is no, but I have never considered the potential integration of the two.
3. I went to a Catholic university and saw how the requirements in religion/philosophy were impactful across the disciplines. I think something like Catholic Social Teaching, which has such a rich history, is very helpful when combined with an overall programme in ethics/philosophy [PI].
4. Our product development has begun to include a range that supports spiritual practice. This could appeal to religious/faith-based groups such as Buddhist, Hindu, or other [PI].
5. Yes—with regard to general principles of behaviour but it would take a lot of companies acting together to influence the world economy [PI]. However, I suppose that has been happening over the past 10 years or so in the area of ethics.
6. However, the challenge business has is to ensure that we are not putting our reputation at risk from organisations who disagree [RK]. As a business, we need to be seen to be apolitical and non-religious and this means at times good partners may be ruled out because of governance checks.
7. Undecided.
8. As an atheist, I would struggle to see the benefit of Business Development [NI].
9. The only link I can see would be indirect for staff involved in such an organisation.
10. Just do the right thing and you will not be too far out. A quote from Brian Cody Kilkenny hurling manager at a breakfast briefing I attended.
11. As long as belief in a certain religion is never a requirement for participation in a business environment, the value of exposure to religious social teaching cannot be undervalued [PI]. It is the tenets of the Judeo-Christian religion that have actually formed most western thought and action, and in a business context, they can be a powerful force for good, positive social, and environmental behaviour.
12. I think having a strong understanding of different cultures (which may emerge as a result of religious beliefs) can have a positive influence, however, I don't believe that a full module should be dedicated to 'Religious Social Teaching' rather the issue covered as an aspect of a different module.
13. It could be positive or negative [RK].
14. I can't make an immediate link between the two.
15. Cannot understand this question.

16. I'm not sure of the relevancy of this question.
17. Business ethics are a focus area, but I wouldn't necessarily link them to religious basis.
18. N/A

Codes:

[NI] No Impact (1/17)
[RK] Risk (3/17)
[PI] Positive Impact (4/17)

Appendix 7D

Question 14: Comment on the level of support you receive in your CSR/Sustainable Development duties within your organisation.

Level of support:

- A small group is very supportive, including the CEO [TM]. However, the vast majority of the company does not see CSR as a primary objective. Rather increasing profits [SP] and meeting business objectives. It is only now that sustainability-related objectives are being added to the short-term goals of the company. This should help.
- Very strong by subject experts needed to complete reporting requirements.
- Sustainability is not a particularly high priority for the organisation [LS].
- Very good support.
- Supported by the Board to the CEO [TM] on down. It pervades the culture. (nice place to be, by the way).
- Highly supportive.
- Support provided through the executive level [TM].
- Medium.
- Not a great deal [LS].
- Full support from CEO [TM] level down.
- I receive full support from the senior management [TM].
- There is a now corporate belief that a commercial business is a sustainable business but sustainability can still be seen by some as an add-on and an unnecessary hindrance to getting on with making the money. However, I think this is now more of a perception and it is associated with the word 'sustainability' as the business does genuinely try to act in a sustainable manner as part of everyday business. To answer the question, therefore, support is generally available if the sustainability team can be seen to be adding value [SP] and not just introducing something that does not add anything to the business.
- Huge amounts of support—It matters a great deal to the business leaders [TM].
- High-level support—from Executive and Board level [TM] as well as the operational teams who run the business in a responsible way every day.
- Good.
- I am supported 100%.
- We have the full support from the Board and the Executive [TM].

- Our whole organisation is setup in a sustainable and responsible way, therefore, there is lots of support. It isn't seen as something separate, just the way we do business.
- Steering Committee which meets 5 times per year and is made up of senior managers [TM] across the business. Report to Executive and Board level.
- The support is there we just have to have a good business case [SP].
- Top-down leadership [TM]: support for strategy—resources, freedom to define and deliver best practice.
- High-level support and oversight by executive [TM] and non-executive.
- High—as a waste management company it is what we do.
- Good support. It is very well understood as part of our business strategy and receives strong support from our senior leadership [TM].
- Personally I co-ordinate CR within our organisation and have a representative in each office who manages CR at a regional level. This is supported by top-level management [TM], with Managing Partners having accountability for our CR and Environmental Sustainability. Our CSR incorporates our staff, suppliers, communities, environment, and customers.
- Good management support.
- None [LS].
- I have sole responsibility, no support [LS].
- Every increasing from all levels from director [TM] to shop floor levels.
- We work well as a team.
- All managers and team leaders are part of our sustainability programme and we strive to include general operatives in areas relevant to them.
- Initially it was limited, but once you start to generate positivity among staff and achieve verifiable resource reductions [SP], the support increased.
- The more sense it makes, the more support it gets.
- Top-level management [TM] is very supportive of our environmental agenda in that there is no issue with capital expenditure to facilitate environmental measures. Our biggest issue would be with staffing levels to provide support for the sustainability measures. Our personnel who meet to implement the sustainability measures are often so busy; it is hard to make time to continue to innovate, solve, measure, and record environmental issues.
- Varied across departments—production engaged and driving sustainability activities, less so in commercial departments.
- It's a growing area, recognised within the business as required. My organisation has recently created a Head of Sustainability role to lead this agenda.
- Little to none [LS].
- As a owner, I decide this priority myself.
- Again, as we are starting out, we are still developing the framework for our CSR activities; this is driven by the MD [TM], so support is strong.
- CSR is driven by top-level management [TM].
- Full support. Sustainability is essential for our company.
- 100% commitment from the owners of the organisation [TM].
- The company is fully supportive and have a strong CSR strategy

- As I am the MD/CEO [TM]—full support
- I receive support from colleges and management, and also from Bord Bia.
- High-level support from Senior management [TM].
- Direct reporting line to executive with strong CEO buy-in [TM]. In the organisation ,we believe that CSR/Sustainability is everyone's role and therefore rather than expanding a dedicated CSR team, we incorporate CSR into every team and so receive support from across all areas of the business.
- The company as part of the mission statement is fully committed to being involved in all aspects of CRS strategy [TM] and as such gets huge support from all involved.
- Low [LS].
- As we are an organic company, our levels are very high.
- Induction training and continuous training in social and sustainability keeps all employees up to date with their and the company's responsibilities. Policies and procedures are in place such as ethics and dignity at work. Employees at all levels of the company [TM] are aware of what is expected from management and this makes my job easier in leading. Monthly reporting is in place, and we are subject to annual audits on social accountability and environmental compliance.
- We have support from Bord Bia, Leo, etc. in regard to this. We also have support from a third party individual who is helping on a voluntary basis to keep themselves up to date with these accreditations.
- Full support, regular meetings, and planning.
- Good support.

Codes:

[TM] Top Management (23/54)
[SP] Support if it is Profitable (3/54)
[LS] Low level of Support (6/54)

Appendix 7E
Question 15: Feel free to comment on this issue about governmental involvement (19% of respondents commented on this question).

1. I feel this question is too broad. It should be decided by each country what is needed and when.
2. Governments can and should legislate [GR] to prevent the most negative impacts a company may have on society, e.g. environmental, minimum wage agreements, banning zero hours contracts, etc. In my opinion, however, CSR should be authentic, and come from the desire of the leaders of a company to do business responsibly. If this is not authentic, then no amount of Government legislation will make up for that.
3. Encouraging, yes—I'm not sure about having more influence [CR] for promotional purposes—Could that become similar to lobbying?
4. Yes—many businesses I speak to are waiting for things to become 'necessary' [GR] and think it is someone else's responsibility. Although I think businesses should take responsibility for the sustainable delivery of their operations, sadly

this may mean it takes longer to see a significant shift in the norm and this could be improved by stronger leadership from government.

5. They should help to set the minimum standards expected and provide avenues/grants/regulation [GR] to encourage leadership and innovation
6. Businesses won't do it on their own [GR].
7. They should lead from the front [GR], 'do as I do not as I say'.
8. The Origin Green programme is the perfect example of the right balance of government and private sector [PP] tackling the issue. If you want to reap the benefits of the Bord Bia marketing system, you have to comply. It's a good trade off. I think that our basic environmental protection measures need to be legislated for [GR] but that sustainability by its nature should not have to be actually legislated for. It is simply the smart thing to do for your businesses in terms of protecting your resources, being efficient and marketing yourself B2B and to consumers [CR].
9. Governments could do more to support this area in business [GR]. In my view, some business is ahead of government in this area and has taken a leadership role.
10. Yes [GR], but it should be more carrot and less stick. Many businesses are so far from sustainable, or the possibility of being sustainable is almost 0% due to the sectors they function in, but that is all the more reason to try.
11. Sustainable Business Development should be a careful balance of Public–Private Partnership [PP]. Governments can be subject to certain levels of lobbying from large industry and so there should be checks and balances from both sides. A relevant example of this is in the Origin Green certification process. It allows large-scale industrial meat industries to achieve the same level of certification as another company that has sustainability as a core business philosophy, while the large businesses do not live up to what the certification stands for. This leads to a dilution of its value.
12. I think that Ireland's inherent green image could be further leveraged and that the government has a role in promoting sustainable business development [GR]. The DJEI (Department of Jobs, Enterprise, and Innovation) have a number of plans (CSR andDelivering Our Green Potential as well as the Action Plan for Jobs) in place, however, the results of these plans remain to be seen.

Codes:

[GR] Governmental Responsibility (8/12)
[PP] Public/Private Partnership (2/12)
[CR] Corporate Responsibility (2/12)

Appendix 7F

Question 16: In your view, what are the major barriers preventing CSR/Sustainable Development managers from leading their businesses up to a higher sustainability level as suggested in sect. 7.4.2.5?

1. Non-partisan leadership [LL] who can bring businesses together to see sustainable development as an URGENT need for all. Companies are still too short-term focused on making money [PD]. Need incentives beyond financial that push corporations to make changes

2. I don't recall Q5 specifically, but major progress would result from a simpler, more compact set of metrics [DM] and proof points. We're trying to boil the ocean

3. Lack of organisational buy-in [LL], lack of clear goals, lack of investor priority/demand [NS]

4. Lack of interest in the form of specific requests for information or questions from key external stakeholders [NS]

5. There is still confusion as to the value of CR [DM] (although that's lessening). So there is a need to be able to tie that value directly to the business and to build relationships across an organisation in all disciplines/divisions. If someone is not well skilled either in the business or in the ability to network and build relationships that will be a large barrier to success

6. Mandates; cost; is it what customers want? Increasingly yes…but they have to be willing to pay for it [NS]?

7. Lack of understanding of the breadth of sustainability [ED]

8. Commercial pressures from customers [EX] (not willing to pay for a more sustainable product) and suppliers (mainly consultants, rating agencies, and others creating ever more work for sustainability professionals)

9. Businesses are ultimately answerable to shareholders [PD], and their focus is not on CSR. The caveat is a situation where a business does not operate responsibly and this leads to financial loss. In my view, however, even in this situation, shareholders only want to feel comfortable that enough has been done within the company to manage risk—I do not think the majority of shareholders feel strongly about whether or not a business operates responsibly [NS]

10. Lack of funds [EX]

11. A perception that 'sustainability' is expensive [EX] and does not add value [ED]. We need to talk about business resilience instead

12. For us, it's a cost thing [EX]—are people willing to invest and think more long-term [NS]? If a sustainable option is the same cost or less, it's a given, if it's more costly, it's always harder to guarantee

13. Resource [EX], focus, and understanding [ED]

14. Cost [EX]

15. I don't think there are any barriers, CSR work, in my experience, is all positive and it baffles me that we do not hear more positive experiences from Senior Management

16. Perceived barriers include budget [EX], resourcing [TM], and conflict with other priorities. I say perceived as I believe it makes business sense to make these changes and will result in savings and/or more business in the long run

17. Sustainability is a very future-driven mission, with minimum short-term gain/value to the business. Managers who concentrate on short term (often sales [PD] or product portfolio) are hard to convince [NS] of the merits

18. Their message/USP needs to resonate with business focused executives [NS], i.e. how can we add value from CSR

19. Lack of vision and foresight Lack of more stringent regulation [LG]

20. None within the company

(continued)

(continued)

21. Speaking the language of the business. It is crucial for CSR managers to work closely with different functions to embed sustainability within their strategies, rather than seeing themselves as separate (often in corporate affairs)

22. lack of senior management buy-in [LL]

23. Education [ED]

24. Time [TM], culture of organisation, ethos of organisation, in a busy work environment, with daily challenges, it can be difficult to focus on sustainability [NS]

25. Short tenure of Board members of companies resulting in a lack of medium to long-term sustainable business strategy [NS]. Lack of clear leadership [LL], vision, mission, and non-aligned values

26. We are a small company with management having enough to do on a day to day basis [TM] [NS]

27. Time [TM]/resource [EX] pressure on managers in business [NS], only the large companies can have people dedicated to CSR

28. Not all senior management buy into CSR [LL] [NS], still come up against old school management style of continuously pushing the bottom line [PD]

29. For my company working with old buildings and equipment without financial resources [EX], it is difficult to do all you want to do

30. I think the biggest stumbling block in our case is that there is no dedicated CSR manager with the time [TM] and energy to dedicate to the task. It is only one element of my role at present as we are an SME. Also, trying to get non CSR managers to take the time to innovate and test new ways of doing things can prove difficult in SMEs that operate on lean staffing levels. There would also be a certain fear factor in trying to gauge how much our customers (retailers) and consumers are willing to give in terms of price hikes (to fund activities) and quality issues (less packaging, lighter materials) as well

31. Lack of metrics to measure the impact of activities [DM] across all areas, including social

32. Perception [ED] that CSR is a cost [EX] to the business, difficulty in measuring the tangible value [DM] it delivers

33. In terms of mission statement level, and failing to have sustainability embedded as part of the mission, I don't believe that limited liability companies can achieve this due to an inherent contradiction between directors' duties in terms of generating profit [PD], and many sustainability and CSR goals. In the case of publically listed companies, the demands of shareholders to generate returns mean that CSR and sustainability projects can only be advanced where there is a clear financial benefit resulting [NS]. Putting CSR/sustainability into the mission statement is either green-washing, or going to affect financial performance. Therefore, government [LG] needs to adjust the playing field (for instance more carbon taxes, penalties for pollution enforced, etc.) to make the sustainability investments within companies financially justifiable

34. Lack of resources [EX] and lack of employee engagement [NS]

35. Lack of sustainability culture in organisation. Lack of capital investment [EX]

36. Money and resources [EX]

37. Cost [EX]

38. As a small business not having an assigned CSR Manager is a barrier and as mentioned in an earlier answer, time is precious

(continued)

(continued)

39. Finance is the obvious one [EX]. As I mentioned, lack of support from social agencies/charities can also be a barrier. If companies don't know what is out there or where their help would best fit, then this is a barrier. Education is king [ED]!

40. The cost that is involved in the installation of equipment [EX]

41. Resources [EX]

42. It's not yet seen as a core issue for most business [NS]. There needs to be a stronger focus on building a business case for CSR and showing hard tangible benefits [DM]. In my opinion, the governments green public procurement will (eventually) be a major driving force as it will illustrate the need to be green in order to win business

43. Understanding fully and committing to what CSR really means [ED] and not just paying lip-service

44. Not enough time [TM]

45. The financial barriers as often the sustainable way is often quite expensive [EX] to implement or slightly more expensive than the alternative, it is also more time [TM] expensive

46. We are a food business with a lot of audits and standards to comply with. Barriers include lack of information from government bodies, local councils, and funding available [EX]. No grants available, no programmes such as STEM in the UK [ED] to get companies started

47. Time [TM], Money [EX], Knowledge [ED]

48. Origin Green is the only real platform Irish Businesses have for a formal development programme, before this programme, there was a lack of support for small businesses

49. Supply chain collaboration

50. Not a clear image of commercial winnings [DM]

Codes:

LL—Lack of Leadership (5/50)
PD—Profit Driven (4/50)
ED— Education (8/50)
TM— Time Resource (8/50)
DM—Difficult to Measure (6/50)
EX— CSR is Expensive (20/50)
LG—Lack of Governance (1/50)
NS—No Stakeholder Priority (15/50)

Appendix 7G

Question 17: If you wish, feel free to add your comments on any pertinent issues arising from, or in addition to, the questions in this survey?

1. CEO commitment to the responsible business agenda is so important [TM]. Without this CSR, managers will struggle to make changes within a company. A CEO with a vision for how the company can make profit while operating responsibly is inspiring

2. Sustainability has to be flexible [FX] with changing situations

(continued)

(continued)

| 3. While I was educated to master's level, it was not a business degree, but rather horticulture [ED], so the questions related to what modules I took in my education may not be so relevant |
| 4. None. In our company, we used the STEM project to get the environment plans started many years ago, this lead to the ISO 14001. Bord Bia lead workshops [ED] in sustainability for companies to work towards certification |

Codes:

[TM] Top Management Commitment
[ED] Education
[FX] Flexible

Appendix 7H

A sample of corporate website content is reproduced (anonymously) for four of the 86 businesses that responded to the survey presented in Chapter 7. A contextual content analysis method of inquiry (McTavish and Pirro 1990) is carried out to determine the contextual information contained in the company text:

- Respondent #35 is a top manager employed in a small (between 10 and 49 employees) family-based dairy products manufacturer based in Ireland who operate on the middle level of CSR activities and engagement with the overall business strategy. The respondent has between 1 and 5 years' experience in a CSR role and the company have the following information on their website: 'We can't change the world, but we can change our small corner...At (Company Name), our team is dedicated to clean simple food bringing authentic farmhouse taste from our farm to your table with honesty and transparency. Subconsciously we were brought up this way, it was normal to practice sustainability and social responsibility. It is our mission to carry them on, to value and hold dear what's good and right. Like never before we are conscious of depleting resources. We are aware we need to change and are learning how to do it. Our diverse sustainability projects are innovative and exciting: Solar energy production, heat recovery, and rainwater harvesting are simple but effective. We are proud of our staff and local community. Our rural heritage is precious to us, and we want to protect it for our children's children'.

- Respondent #70 is a middle manager working in the CSR area for a period between 1 and 5 years with a UK-based property solutions (construction and development) company with a total number of employees between 50 and 249 employees. The respondent claims that they are operating at the upper level of CSR integration and the main disadvantages with CSR reporting are collating the information and measuring the impact of the report. They have a CR (Corporate Responsibility) tab on their website homepage and this link states the following information: 'Corporate responsibility at (Company Name) is about delivering great projects while upholding a clear commitment to our values'. The company vision and values are then described as 'After 70 years, the key parts of our DNA as a business can still be seen'. We don't just believe business is a financial, technical, or logical transaction, we believe we are making a connection between people'.

- Respondent #68 is a Master's level educated top manager employed by a global South-African based financial institution with a workforce greater than 100,000 employees. The respondent claims that they are operating at the upper level of CSR integration and the company sustainability efforts are based on the following plan as stated on their company website: 'On a global and national scale, there are many social, environmental, and economic challenges and opportunities that will face us as a business over the next 10 to 20 years. By addressing these challenges, and turning them into opportunities, we will create mutual benefit for the societies in which we operate our business in the long-term'. Some of these large-scale pressures and interconnected trends include

  - Large-scale unemployment, particularly amongst youth. This is particularly pertinent across Africa where 40% of the world's youth will be seeking jobs and a means of making a living and where approximately 50% of the population is financially excluded.
  - Climate change and extreme weather patterns. These are disrupting agricultural systems, resulting in food scarcity, a rise in food prices and pushing people into poverty.
  - An ageing and increasing urban population. This is putting pressure on natural resources, education systems, housing, and infrastructure. More investment is needed to provide for an ageing population, particularly in health and financial management.
  - Transformation to digitised economies which will open up opportunities for many in sharing information and accessing services. However, could also exacerbate income differences when populations do not quickly adapt and benefit from technology-driven growth. Our Plan aims to contribute to the solutions to some of these challenges while also aligning to international and national plans. In particular, a number of our programmes align with the United Nations Sustainable Development Goals (SDGs), and we support the South African Government's National Development Plan (NDP). These frameworks help us as we develop cross-sector partnerships to address common issues. As we progress with our Plan, we will be able to demonstrate how our activities positively contribute to national and global issues.

- Respondent #85 is a Master's level educated top manager who works with a US-based utility company with between 10,000 and 100,000 employees, and the company operates at the upper CSR engagement level. The respondent states that the level of support for CSR is very strong by subject experts needed to complete reporting requirements and has the following information on their website: '(Company Name) continues to be recognised as a trailblazer in environmental stewardship and corporate sustainability. We were the first U.S. utility to voluntarily commit to stabilising CO_2 emissions, and for 14 consecutive years, the Dow Jones Sustainability Index has included (Company Name) on either its World or North America index or both. Developing solutions for the environment requires short- and long-term actions along with commitment and follow-through.

In 2011, we adopted Environment 2020, a comprehensive environmental strategy and management system that covers six areas of strategic action'.

Analysis of the website content from a global sample of the survey respondents (#35—Irish, #70—the UK, #86—South African, #85—US) demonstrates the focus by business to highlight the non-financial dimensions (human and environmental) of their business efforts and activities. The financial aspect is not mentioned in any content. It should be noted that many organisations are driven by their desire to fit in with their external environment (Biloslava and Lynn 2007) and the aforementioned company statements are examples of such efforts. Respondent #35 is different from the other three examples in that it is a small family-based business and shows a strong relative emphasis on both the *traditional* and *emotional* context dimensions as described by McTavish and Pirro (1990). Indeed Campopiano and De Massis (2015) claim that family firms show several differences in the type and content of CSR reports than non-family firms. Respondent #68 has an emphasis on the *practical* context dimension. Respondent #85 emphasises the *analytical* context dimension and as a non-family firm are keener to communicate compliance to CSR standards than family-run firms (Campopiano and De Massis 2015). Respondent #70 emphasises the *traditional* context dimension (McTavish and Pirro 1990). It is clear from the contextual analysis on website content from Respondents #35, #70, #68, and #85 that consumers are the main focus when content is loaded onto the company website. Businesses want to portray themselves as being responsible, caring entities and, as part of the business marketing strategy, have an opportunity to insert company content on the website that can contribute to this positive goal. However, questions remain as to whether the increased amount of CSR communications by businesses has led to an improved goodwill factor on the side of stakeholders (Lock and Seele 2016).

Appendix 7I

Origin Green Sustainability Reporting—Overview
Origin Green is an initiative of Bord Bia (2015), the Irish food board and was first launched in 2012. It is focused on the food and drink sector, and it claims to assist farmers and producers to set and achieve measurable sustainability targets. The main target area includes reducing the environmental impact, serving local communities, and protecting the natural resources in Ireland. As a sample of the initiatives being undertaken by businesses, Respondent #41 is an Irish food producer, and on its sustainability plan, the following elements are included:

- Facilities/Capital initiatives (Wind Turbine, Water Harvesting, Capital Energy investments),
- Purchasing initiatives (local sourcing of products, encouraging awareness of sustainability issues by supply chain),
- Production/Logistics initiatives (improving production processes, better management of resources),
- Community initiatives (healthy eating programmes, community food network participation, local employment initiatives, summer intern programmes).

Origin Green claims that 512 companies have so far signed up to the Origin Green Programme (accessed on 10 August 2016). The Irish food and drink industry is heavily dependent on export and the Origin Green accreditation, independently verified, is seen as an important marketing tool.

Global Reporting Initiative (GRI) Sustainability Reporting—Overview

Global Reporting Initiative (GRI 2017) is an international, independent standards organisation that helps businesses, governments, and other organisations understand and communicate the impact of business on critical sustainability issues such as climate change, human rights, corruption, and many others. GRI was founded in Boston in 1997. Its roots lay in the US non-profit organisations, the Coalition for Environmentally Responsible Economies (CERES) and the Tellus Institute (Tellus Institute 2017). The Global Reporting Initiative (GRI) produces standards for sustainability reporting, the GRI Guidelines. The GRI claim that of the world's largest 250 corporations, 93% report on their sustainability performance and 82% of these corporations use the GRI standards to do so. The most up-to-date GRI guidelines at present (August 2016) are the G4 Guidelines. G4 is designed to be universally applicable to all organisations of all types and sectors, large and small, across the world. When a company decides to voluntarily report on their performance using the GRI (G4) framework, they must complete a comprehensive document with a view of the wider context of sustainability. In terms of material included in the report, it must (i) reflect the organisation's three significant categories namely economic, environmental, and social impacts (ii) substantively influence the assessments and decisions of stakeholders. The social category is further sub-divided into (i) labour practices and decent work (ii) human rights (iii) society (iv) product responsibility. A very detailed 269-page implementation manual is provided from the GRI website to assist companies in completing the report. The report is completed on an annual basis.

References

Biloslava, R., & Lynn, M. (2007). Mission statement in slovene enterprises, institutional pressures and contextual adaptation. *Management Decision, 45*(4), 773–788. https://doi.org/10.1108/002517407/10746024.

Bord Bia (2015). *Irish Food Board*. Available at http://www.bordbia.ie. Accessed 6 November 2017.

Campopiano, G., & De Massis, A. (2015). Corporate social responsibility reporting: A content analysis in family and non-family firms. *Journal of Business Ethics, 129*(3), 511–534. https://doi.org/10.1007/s.10551- 014-2174-z.

Commission for Energy Regulation, Electricity and Gas Retail Markets Annual Report 2013, Information Paper, CER/14/134, published 25th June 2014.

Department of Jobs, Enterprise, and Innovation (DJEI). (2017). *'Towards responsible business', Ireland's National Plan on Corporate Social Responsibility 2017–2020*. Available at https://www.djei.ie/. Accessed on 14 August 2017.

Environmental Protection Agency (EPA). Available at http://epa.ie/. Accessed on 6 September 2017.

Eurostat, Statistical Office of the European Union. Available at http://ec.europa.eu/eurostat. Accessed on 10 February 2016.

Gorta. Available at http://www.gorta.org. Accessed on 24 February 2014.

Global Reporting Initiative (GRI). (2000). *Sustainability reporting guidelines on economic, environmental, and social performance*. Boston: Global Reporting Initiative.

Global Reporting Initiative (GRI). (2017a). *Materiality in the context of the GRI reporting framework*. Available at http://www.globalreporting.org. Accessed on 27 November 2017.

Global Reporting Initiative (GRI). (2017b). *Global Reporting initiative*. Available at http://www.globalreporting.org. Accessed on 6 November 2017.

Global Reporting Initiative (GRI). *Reports list*. Available at https://www.globalreporting.org/services/Analysis/Reports_List/Pages/default.aspx. Accessed on 28 April 2017.

Horizon 2020, The EU Framework Programme for Research and Innovation. Available at https://ec.europa.eu/programmes/horizon2020//en. Accessed on 3 March 2017.

Innovation 2020: DJEI, Department of Jobs, Enterprise and Innovation. Available at http://www.djei.ie/en/Publications/Innovation-2020.html. Accessed on 3 March 2017.

ISO 26000 Guidance Standard on Social Responsibility. Available at http://www.iso.org/iso/home/standards/iso26000.htm. Accessed on 2 March 2017.

Lock, I., & Seele, P. (2016, May). The credibility of CSR (corporate social responsibility) reports in Europe. Evidence from a quantitative content analysis in 11 countries. *Journal of Cleaner Production, 122*, 186–200. https://doi.org/10.1016/j.jclepro.2016.02.060.

McTavish, D. G., & Pirro, E. B. (1990). Contextual content analysis. *Quality & Quantity, 24*, 245–265. Kluwer Academic Publishers.

Origin Green, Bord Bia. Available at http://www.origingreen.ie. Accessed on 16 June 2014.

© Springer Nature Switzerland AG 2020

T. Kealy, *Evaluating Sustainable Development and Corporate Social Responsibility Projects*, https://doi.org/10.1007/978-3-030-38673-3

Origin Green, Verified Members. Available at http://www.origingreen.ie/companies/verified-members/. Accessed on 28 April 2017.

Tellus Institute. (2017). Available at http://www.tellus.org. Accessed on 6 November 2017.

United Nations Global Compact. Available at http://www.unglobalcompact.org/aboutthegc/thetenprinciples/. Accessed on 2 March 2017.

Printed by Printforce, the Netherlands